PC-DSP

$5\frac{1}{4}$" IBM® Version

Oktay Alkin

Southern Illinois University at Edwardsville

PRENTICE HALL, Englewood Cliffs, New Jersey 07632

Library of Congress Cataloging-in-Publication Data

ALKIN, OKTAY

PC-DSP, $5\frac{1}{4}$" IBM version / Oktay Alkin.

p. cm.

Includes bibliographical references.

ISBN 0–13–655200–5

1. Signal processing--Digital techniques--Data processing.
2. Microcomputers. I. Title.

TK5102.5.A427 1990

621.382'2--dc20 89–37044
CIP

Editorial/production supervision: *Edith Riker/DebraAnn Thompson*
Cover design: *Ben Santora*
Manufacturing buyer: *Margaret Rizzi*

Printed in the United States of America

10 9 8 7 6 5 4 3 2 1

ISBN 0-13-655200-5

Prentice-Hall International (UK) Limited, *London*
Prentice-Hall of Australia Pty. Limited, *Sydney*
Prentice-Hall Canada Inc., *Toronto*
Prentice-Hall Hispanoamericana, S.A., *Mexico*
Prentice-Hall of India Private Limited, *New Delhi*
Prentice-Hall of Japan, Inc., *Tokyo*
Simon & Schuster Asia Pte. Ltd., *Singapore*
Editora Prentice-Hall do Brasil, Ltda., *Rio de Janeiro*

To Esin, Sinan, and Deniz

Contents

Preface

PC-DSP (Personal Computer–Digital Signal Processing) is an interactive, menu-driven software package designed for use in performing common digital signal processing tasks such as waveform synthesis and manipulation, convolution, correlation, fast Fourier transforms, z-transform, design and analysis of digital filters, filtering of discrete data, and estimation of power spectral density. Although it was primarily designed for use as a supplement in a senior-level or graduate course in digital signal processing, it can also be a valuable tool for the practicing engineer who wishes to enhance her understanding of this vital subject.

Today, there are a number of excellent textbooks available in the area of digital signal processing. Nevertheless, one shortcoming that is often overlooked is the lack of proper software tools for the student to use in applying gained theoretical knowledge to realistic signal processing problems. Any signal processing software that is readily available to the student through textbooks is often in the form of source listings of several routines provided in an appendix or sometimes on an accompanying computer diskette. In most cases, each routine is written to perform one specific task, and the student is expected to type in (or at least compile) these routines before they can be put to use. Commonly available commercial software packages such as spreadsheets or equation-solvers can close this gap to a certain extent, but the student usually ends up spending more time than is necessary, owing to the fact that he is trying to use a software product for something other than its intended purpose. Also, the time spent in this way is not very productive, since most of it is spent trying to figure out how to use the software rather than concentrating on the problem itself.

PC-DSP was written to provide the student with an integrated signal processing environment in which realistic digital signal processing problems can be solved. It employs an interactive user interface. No programming knowledge is required on the part of the user, although basic familiarity with personal computers and the disk operating system (DOS) is helpful. One can generate, analyze, and process data by going through a set of menu choices, without the need to remember any command syntax. A context-sensitive on-line help feature is also provided for added ease of use. Real or complex discrete data sequences can either be internally generated or loaded from disk data files. The results of PC-DSP operations can be tabulated, plotted, or written into disk data files. PC-DSP provides internal memory storage for up to 50 named complex sequences (up to 16,000 real data values) to permit the solution of relatively complicated problems.

The program will work on an IBM or fully compatible personal computer operating under DOS version 3.0 or higher, with at least one disk drive, 512 k-bytes of random access memory (RAM), and a Color Graphics Adapter (CGA). If an Enhanced Graphics Adapter (EGA) is available, the program will take advantage of it to produce higher-quality text and graphics. It will also take advantage of a math co-processor chip (if one is installed) to speed-up floating point math operations.

Options are provided for generating uniform or Gaussian random noise, trigonometric and exponential sequences, discrete window functions, and linear ramp squences. Basic arithmetic and trigonometric operations such as addition, multiplication, sine, cosine, logarithm, and exponential can be performed on complex sequences. Cut-and-paste editing functions allow the user to generate sequences with more complicated definitions.

Finite-Impulse-Response (FIR), and Infinite-Impulse-Response (IIR) digital filters can be designed and/or analyzed using various techniques. FIR filter design can be accomplished by the use of Fourier series method with windows, Parks-McClellan algorithm, or the least-squares design method. Butterworth and Chebyshev IIR filters can be designed using the bilinear transformation method. Designed filters can be analyzed by means of impulse response, pole-zero plots, magnitude, log-magnitude, phase, and group delay characteristics. Discrete data sequences can be processed through designed filters.

Spectrum analysis can be performed using classical methods such as the periodogram (or its variation due to Welch) and the Blackman-Tukey method, as well as modern spectrum analysis methods that utilize AR, MA, and ARMA modeling techniques.

New functions can be created out of the existing ones using the macro file capability. This feature also makes it possible to interface user-written executable files to PC-DSP, thus making them part of an integrated signal processing environment.

It is the author's opinion that mathematical results often provide a better insight into the problem at hand if they can be viewed graphically. Therefore, a set of sophisticated graphics routines have been developed to enable the user to "look at" the results of signal processing operations with minimal effort. The results of most operations can be graphed instantly with one key. Graphs can also be customized for use in formal reports. One or two sequences can be plotted at a time. The user has complete control of graph parameters such as ranges, title, axis labels, increments, and the like. Plots can either be continuous or discrete. They can be viewed on the screen, printed on a dot-matrix printer with graphics capability, or saved into a disk file for later use.

This manual/workbook consists of nine chapters and two appendices. It does not attempt to teach digital signal processing. Its primary purpose is to teach the usage of the program, to increase the user's curiosity in signal processing concepts, and to motivate exploration. A good number of solved examples are used in explaining how to use the program. Exercises to be worked by the student are given at the end of appropriate chapters and can be used by instructors as laboratory experiments or computer assignments. Typically, each exercise concentrates on one fundamental digital signal processing concept.

Chapter 1 begins by explaining the required hardware and the contents of program disks. Procedures for making backup copies of program disks and for installing and running PC-DSP on various hardware configurations are also detailed in this chapter. In Chapter 2, the reader is guided through the menu structure and the general features of PC-DSP. Procedures for loading, saving, printing, and deleting discrete sequences are briefly illustrated. Chapter 3 explains the generic waveform synthesis functions available. Types of data files and data file input-output are also discussed in this chapter. In Chapter 4, basic signal processing capabilities of the program such as FFT, DFT, convolution, and correlation are explained through the use of example problems. Graphics capabilities of PC-DSP are exploited in Chapter 5. Methods for customizing graphs are discussed. Graphics data files and dot-matrix printer graphics are explained. Chapter 6 treats the design and analysis of FIR filters. Fundamentals of the three design techniques employed are briefly explained. IIR filter analysis and design is discussed in Chapter 7. Simulation of designed IIR filters, and analysis of the student's pencil-and-paper designs are also explained. Chapter 8 concentrates on spectrum estimation applications of PC-DSP. Available spectrum analysis methods are illustrated using the two example data sequences supplied. In Chapter 9, examples of more advanced applications of PC-DSP are given. Macro file capability is illustrated in this chapter. Source listings of sample programs in FORTRAN, BASIC, and PASCAL languages are given to illustrate the procedures for writing add-on code. Appendix A is a function reference list that contains brief descriptions of all PC-DSP functions. Restrictions on parameter values and a brief explanation of the method used by PC-DSP is given where applicable. Appendix B discusses how PC-DSP was developed and how it is being utilized in a senior-level signal processing laboratory at Southern Illinois University—Edwardsville.

ACKNOWLEDGEMENTS

I would like to express my appreciation to those who contributed to the successful completion of the project. The following reviewers have provided very helpful comments and suggestions: Dr. Bernard Hutchins, Cornell University; Dr. William Tranter, University of Missouri-Rolla; John Buck, Kevin Clark, and Rosalind Wright, Massachusetts Institute of Technology. Thanks are also due to Larry Foster who has provided valuable comments and constructive criticism while helping me develop the laboratory sessions for our DSP course. I am also grateful to my students in EE 436 who have made suggestions, pointed out errors in the code, and helped improve the pedagogical aspects of the program. Finally, I would like to thank Tim Bozik of Prentice Hall for this support and enthusiasm throughout the entire project.

1

Getting Started

In this first chapter, minimum hardware requirements for using PC-DSP are specified, and descriptions of the files found in PC-DSP disks are given. The procedures for making backup copies of program disks and for setting up PC-DSP on various hardware configurations are also discussed. At this point resist the temptation that you might have to try out the program right away, at least until you have read the rest of this chapter.

1-1 HARDWARE REQUIREMENTS

In order to use PC-DSP, you will need the following:

- An IBM PC or 100% compatible computer operating under the Microsoft Disk Operating System (MS-DOS) version 3.0 or higher, with at least one 5.25-inch or 3.5-inch disk drive. (Two disk drives—or a hard disk—are recommended).
- At least 512 k-bytes of Random Access Memory (RAM) installed in the system. 640 k-bytes are recommended if user-written programs are to be interfaced to PC-DSP for advanced applications.
- A Color Graphics Adapter (CGA) or an Enhanced Graphics Adapter (EGA) with an appropriate monitor. A nongraphics text display adapter may also be used, but no graphics can be obtained. PC-DSP assumes that you have a CGA by default. If

you have an EGA, or a text display adapter, you must use a command line option to override the default settings. (More on this later.)

- A math coprocessor such as 8087 or 80287 (optional). If it is present in your system, PC-DSP will automatically detect it and use it to speed up calculations.
- A dot-matrix printer is not required but is recommended for obtaining hard copies of tabulations and graphics. To obtain printer graphics, you will also need the proper graphics driver program for your printer (GRAPHICS.COM or similar).
- For proper operation of PC-DSP, it is recommended that the following two lines be placed in the CONFIG.SYS file of the disk operating system:

```
FILES = 20
BUFFERS = 20
```

1-2 CONTENTS OF PC-DSP DISKS

PC-DSP is supplied on either two 5.25-inch floppy disks or one 3.5-inch disk. Both versions contain exactly the same files, the only difference being the magnetic media. In this section and in the rest of this chapter, we will assume that you have the two- disk 5.25-inch floppy version of the program. The first one of the two disks is referred to as the *Program Disk*, and the second is referred to as the *Utilities and Examples Disk*. The contents of these two disks are listed below. If you have the 3.5-inch version, the contents of both floppy disks are stored in one 3.5-inch disk.

PROGRAM DISK

PCDSP.EXE The executable binary program file.

PCDSP.MNU Overlay file containing PC-DSP menu structure and error messages.

DSPMAC.EXE Macro compiler used for writing user macros, and for interfacing other programs to PC-DSP.

UTILITIES AND EXAMPLES DISK

PCDSP.HLP On-line help file. Contains the help information that is displayed on the screen when the [F1] function key is pressed.

PCDSP.MNU Overlay file containing PC-DSP menu structure and error messages.

MYDATA1.DSP A sample data file in binary format.

MYDATA2.DAT A sample data file in ASCII format.

MYGRAPH.GRP A sample graphics data file.

MYMACRO.MAC A sample macro file.

RECTIFY.FOR
RECTIFY.PAS
RECTIFY.BAS } Sample child processes in FORTRAN, PASCAL, and compiled BASIC languages. Used with macro files.

EX???.DSP	Data files used in example problems.
EXER???.DSP	Data files used in exercises.

1-3 BACKING UP YOUR PC-DSP DISKS

Before you use PC-DSP for the first time, you should make work copies of the original program disks. We recommend that you keep the original program disks as backup and use the work copies to run the program. In this way, if you accidentally destroy a disk, you can make another copy from the original. In this section, we will go through the step-by-step procedure for making work copies. Again, we will use the two-disk 5.25-inch floppy version in our discussion, with the understanding that the same applies to the 3.5-inch version.

Now put the Disk Operating System (DOS) disk into drive A, and power-up your system. If DOS prompts you for the date and the time, answer those prompts, as explained in your DOS manual. At this point, you should see the DOS prompt A> on your screen. If so, you are ready to proceed with the backup operation.

For a single-drive system: To begin, you will need two formatted blank disks. If you are not familiar with the procedure of formatting disks, consult your DOS manual.

If you have only one disk drive, you will have to use the DOS command DISKCOPY to make work copies of PC-DSP disks. With the DOS disk placed in drive A, type

```
DISKCOPY A: A:
```

and press the <ENTER> key. DOS will prompt you as follows:

```
Insert SOURCE diskette in drive A:
Press any key when ready . . .
```

Remove the DOS disk from the disk drive, and insert the PC-DSP Program Disk. Press any key. After a few seconds, DOS will prompt you again, this time with the following message:

```
Insert TARGET diskette in drive A:
Press any key when ready . . .
```

Now remove the PC-DSP Program Disk from the drive, and insert one of the formatted blank disks. Again, press any key. When copying is completed, DOS will ask you if you want another copy. Answer this question by typing the letter *y*. DOS will again display the message

```
Insert SOURCE diskette in drive A:
Press any key when ready . . .
```

Remove and label the work copy that you just made. Put the PC-DSP Utilities and Examples Disk in drive A, and press any key. When DOS displays the message

```
Insert TARGET diskette in drive A:
Press any key when ready . . .
```

remove the disk in drive A. Insert the second formatted blank disk, and press any key. This completes the backup procedure.

For a two-drive system: Insert the PC-DSP Program Disk into drive A, and a formatted blank disk into drive B. At the DOS prompt, type

```
COPY A:*.* B:
```

and press the <ENTER> key. This command tells DOS to copy every file on the disk in drive A to the disk in drive B. DOS displays the names of files on the screen as they are copied. When copying is completed, DOS will print the number of files it copied onto the disk in drive B. Remove and label the work copy that you just made. Now insert the PC-DSP Utilities and Examples Disk into drive A and the second formatted blank disk into drive B. Repeat the procedure.

For a hard-disk system: If your computer is equipped with a hard disk, then you will probably want to install the work copy of PC-DSP on the hard disk. We will assume that your hard disk is designated as drive C and DOS has been installed in the root directory of drive C. If these assumptions are correct, then you should see the DOS prompt C> on your screen. (If your hard-disk setup is different from this, consult your DOS manual).

We will start by creating a subdirectory in which PC-DSP files can reside. Do this by typing

```
MD C:\DSP
```

at the DOS prompt. Press the <ENTER> key when finished. This command instructs DOS to create a new subdirectory with the name DSP on drive C. (You may substitute any other name for DSP if you wish). Now put the Program Disk in drive A, type

```
COPY A:*.* C:\DSP\*.*
```

and press <ENTER>. This command tells DOS to copy every file on the disk in drive A to the subdirectory DSP of the hard disk C. File names will be printed on the screen as they are copied. When copying is completed, DOS will print the number of files copied. Now you can remove the original Program Disk from drive A, insert the Utilities and Examples Disk, and repeat the same procedure.

1-4 EXECUTING PC-DSP

Now you are ready to execute PC-DSP. If you have made work copies of PC-DSP on floppy disks, put the Program Disk in drive A, and at the A> prompt, type

```
PCDSP          <ENTER>
```

Note that the word <ENTER> is not meant to be typed in. It just indicates that you should press the <ENTER> key on your keyboard. From this point on, we will use the < ... > notation to indicate the keys to be pressed.

If you made a work copy on your hard disk in the subdirectory C:\DSP, then you will need to make that subdirectory your default directory before you can execute the program. Type the following two commands at the C> prompt:

```
CD C:\DSP      <ENTER>
PCDSP          <ENTER>
```

After performing the preceding steps, you should see the PC-DSP opening message on your screen. By pressing any key or by just waiting for 15 seconds, you can get into the PC-DSP menu structure.

If you are using floppy disks, after the main menu appears at the top of the screen, remove the Program Disk from drive A and insert the Utilities and Examples Disk. (In Sec. 1-6, an alternative method to start PC-DSP will be given for computers with two floppy disk drives.) At this point, we are ready to use PC-DSP menus.

Note: If dot-matrix printer graphics are to be obtained, a graphics driver program such as GRAPHICS.COM, which is in the DOS disk, should be executed before executing PC-DSP. When PC-DSP displays a plot on the screen, a printer copy of the plot can be produced by pressing the <PrtSc> key.

1-5 PC-DSP COMMAND LINE OPTIONS

A command line option is a character preceded by a forward slash (e.g., /E) and is typed following the name of the program at the DOS level. Command line options are used to change the characteristics of the screen display, depending on the graphics adapter and monitor combination available. The default PC-DSP display is white on a black background. A Color Graphics Adapter (CGA) is assumed by default. Three command line options are provided to change the display attributes.

- The /C option produces a color display with regular text shown in white on a blue/green background. The status line and error messages are in yellow with a red background, and help screens are in yellow with a brown background.

- The /E option is used to produce high-resolution graphics (640 by 350 pixels as opposed to 640 by 200 pixel resolution of CGA). You must have an Enhanced Graphics Adapter (EGA) to use this option.
- The /T option forces PC-DSP to bypass graphics routines when they are called. It is included for compatibility with non graphics (text-only) systems and must be used if neither a CGA nor an EGA is available.

The two command line options /C and /E can be used either separately or together but the /T option cannot be used with any of the other two options. Table 1-1 lists five different ways to invoke PC-DSP.

TABLE 1-1

Graphics adapter	Monitor	Type at DOS level
CGA	Monochrome	PCDSP
CGA	Color	PCDSP /C
EGA	Monochrome	PCDSP /E
EGA	Enhanced color	PCDSP /E /C
Text display	Any monitor	PCDSP /T

1-6 PC-DSP ENVIRONMENT VARIABLES

When we executed PC-DSP in Section 1-4, we assumed that all the necessary files were accessible through the same disk drive or, for a hard disk, through the same subdirectory. If you are using floppy disks and have only one disk drive, then all PC-DSP files are indeed accessed through disk drive A, and you may skip this section. Also, if you have a hard disk and wish to place all the files that PC-DSP needs into the same subdirectory, you don't need to be concerned about environment variables.

As your experience with PC-DSP grows, you will probably realize that it is not very convenient to keep all the data files that you have created on the same disk or in the same sub-directory together with PC-DSP program files. If you have two floppy disk drives, you will probably want to use drive A for the Program Disk and drive B for the Utilities and Examples Disk. This can be accomplished using the SET command in DOS to set the values of two environment variables, OVRL and DATA.

When PC-DSP is run, it checks a special location in memory called the DOS environment table to see if these variables are set. The first environment variable, OVRL, specifies the location of overlay files PCDSP.HLP and PCDSP.MNU as well as user-written add-ons with extensions MAC and EXE. The second environment variable, DATA, tells PC-DSP where to search for data files. All files with extensions DSP, DAT, and GRP belong to this category. For a two-drive system, if you type

```
SET OVRL = B:\          <ENTER>
SET DATA = B:\          <ENTER>
PCDSP /options          <ENTER>
```

then you can use drive A: for the Program Disk and drive B: for the Utilities and Examples Disk.

If you have a hard disk and if overlay and data files are to be stored in subdirectories named C:\DSP\OVRL and C:\DSP\DATA, respectively, then type

```
SET OVRL = C:\DSP\OVRL\     <ENTER>
SET DATA = C:\DSP\DATA\     <ENTER>
PCDSP /options              <ENTER>
```

Note that the backslash (\ ) is always the last character in SET command lines. You can also place these commands in a batch file to avoid the need for typing them every time you use PC-DSP. Consult the section on batch files in your DOS manual to find out how to do this.

Another environment variable that is recognized by PC-DSP is TEMP. It specifies the directory location of temporary data files used for data transfer between PC-DSP and user-written child processes (to be covered in Chapter 9).

Some "nearly compatible" computers have been known to hang up whenever a program attempts to use the math coprocessor chip. If this happens on your computer, you can instruct PC-DSP to ignore the coprocessor by using the NO87 environment variable in a SET command, i.e.,

```
SET NO87 = Anything
```

2

Basic Concepts About PC-DSP

PC-DSP is an interactive, menu-driven software package used for analysis, design, and implementation of discrete-time signals and systems. It can be thought of as a special-purpose calculator geared toward performing digital signal processing operations. You will see that it is quite simple and straightforward to use and yet powerful enough to be used by practicing engineers as well as students of digital signal processing. At this point, one word of caution is in order: No computer program eliminates the need for learning the theory of the particular subject matter in hand. PC-DSP is intended to be used as a mathematical tool to enhance the understanding of digital signal processing concepts and to assist in calculations that would be tedious to carry out manually. It is always up to the student, however, to interpret the results correctly and to use them in a meaningful way.

In this chapter, we introduce some of the basic features of PC-DSP and present a brief overview of the program's capabilities. In doing so, we use step-by-step examples to help illustrate PC-DSP menus, special keys, file and variable-handling methods, and the like. When you first start up PC-DSP as explained in Chapter 1, the opening screen shown in Fig. 2-1 is displayed. At this point, pressing any key on the keyboard brings up the main menu, shown in Fig. 2-2. (If you do not press any key, the opening screen is displayed for about 15 seconds, after which the program automatically switches to the main menu). At the top of the screen, there is a menu box with seven menu choices listed in it. The first one of these choices, *System*, is highlighted. This is the main menu of PC-DSP. The highlight can be moved in either direction by using right and left cursor keys on the keyboard. A brief explanation of the highlighted menu item is given on the explanation line toward the bottom of the screen. As you move the highlight with cursor keys, the explanation line changes, depending on the highlighted item. Also note that if you move past the last item in the menu box, the highlight wraps around and enters from the other side.

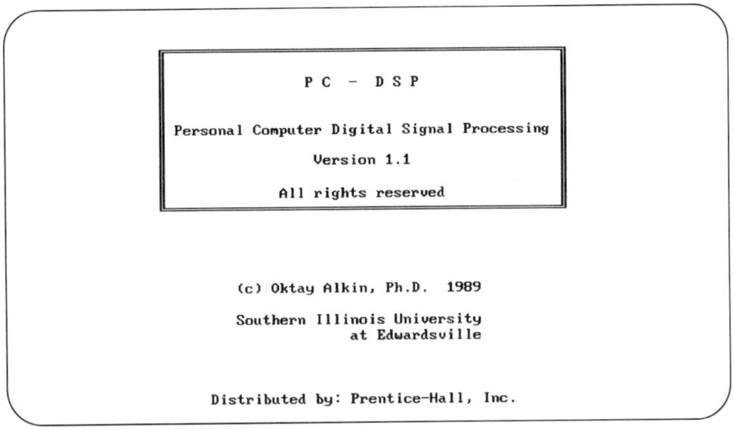

Figure 2-1

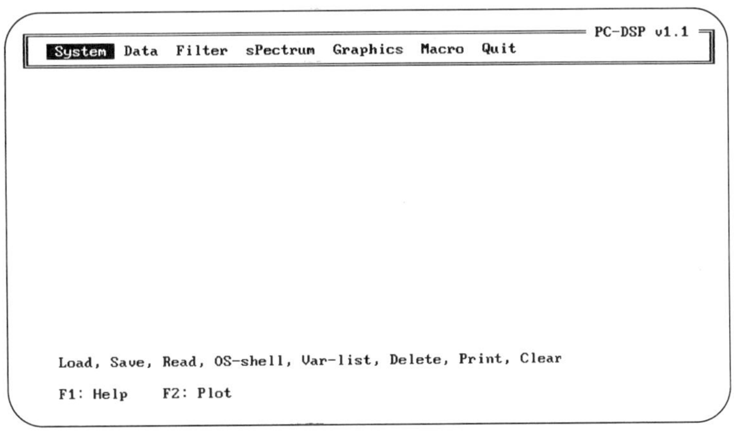

Figure 2-2

The *status line* at the bottom of the screen displays information about special keys, such as function keys, the <ESC> key, and cursor keys. It also displays error messages. As you work in different parts of the program, you will notice that meanings of some keys will change. In particular, the <ESC> key is used for three different purposes throughout the program, and its effect depends on what you were doing when you pressed it. Details of this will be left to the examples.

2-1 PC-DSP MENU STRUCTURE

PC-DSP is built around a set of menus structured in a tree like form. The main menu contains six major components of the program, namely, *System, Data, Filter, sPectrum, Graphics,* and *Macro.* The first five of these six components have their own submenus that are displayed when these components are selected. A complete menu tree diagram is shown in Fig. 2-3. The functions of each major component of PC-DSP will be briefly explained next.

The *System* component is where most of the bookkeeping is done. It is used for things such as writing and reading disk data files, making temporary exits to the operating system, keyboard entry of short data sequences, and for listing, printing, and deleting variables in memory.

Waveform synthesis and basic signal processing take place in the *Data* component of the program, which contains a large variety of functions for generating and manipulating discrete data sequences. Also included in this component are functions to perform most common digital signal processing operations such as Fourier transforms, inverse z-transform, convolution, correlation, decimation, zero-insertion, and zero-padding.

The *Filter* component is used for design and analysis of Finite-Impulse-Response (FIR) and Infinite-Impulse-Response (IIR) digital filters. A variety of design methods and analysis options have been provided. Cascade form realization of designed IIR filters is also done in this component, where arbitrary sequences generated within the *Data* component or loaded from disk files can be processed through designed IIR filters. The modular structure of the program permits analysis and implementation of the user's pencil-and-paper designs as well. FIR processing is left out of the *Filter* component, since it can be done very easily using the convolution function of the *Data* component.

The *sPectrum* component contains some of the classical and modern spectrum analysis algorithms. For a given signal, the power spectral density (PSD) estimate and related parameters can be computed using the Welch method, the Blackman-Tukey method, or time series modeling techniques that utilize autoregressive (AR), moving average (MA), and autoregressive moving average (ARMA) models.

The *Graphics* component is used for generating and displaying customized graphs of discrete sequences. Graphics data file input and output are also accomplished in this part of the program.

The *Macro* component is used for executing special macro files created using the macro compiler DSPMAC. Macro files are made of a series of commands that are executed sequentially. This feature allows the addition of user-defined functions to PC-DSP and also provides a method for interfacing user-written programs, independent of the programming language used.

In PC-DSP, there are two methods for making a selection from any menu. The first method is based on moving the highlight to the desired item using appropriate cursor keys and then pressing the <ENTER> key. An alternative way to make a selection is to type the key letter (first capitalized letter) of the desired menu item. For example, to bring up the *System* menu while in the main menu, you can type the letter *S*. As you start experimenting with the program, you will probably find it a little easier to use the arrow

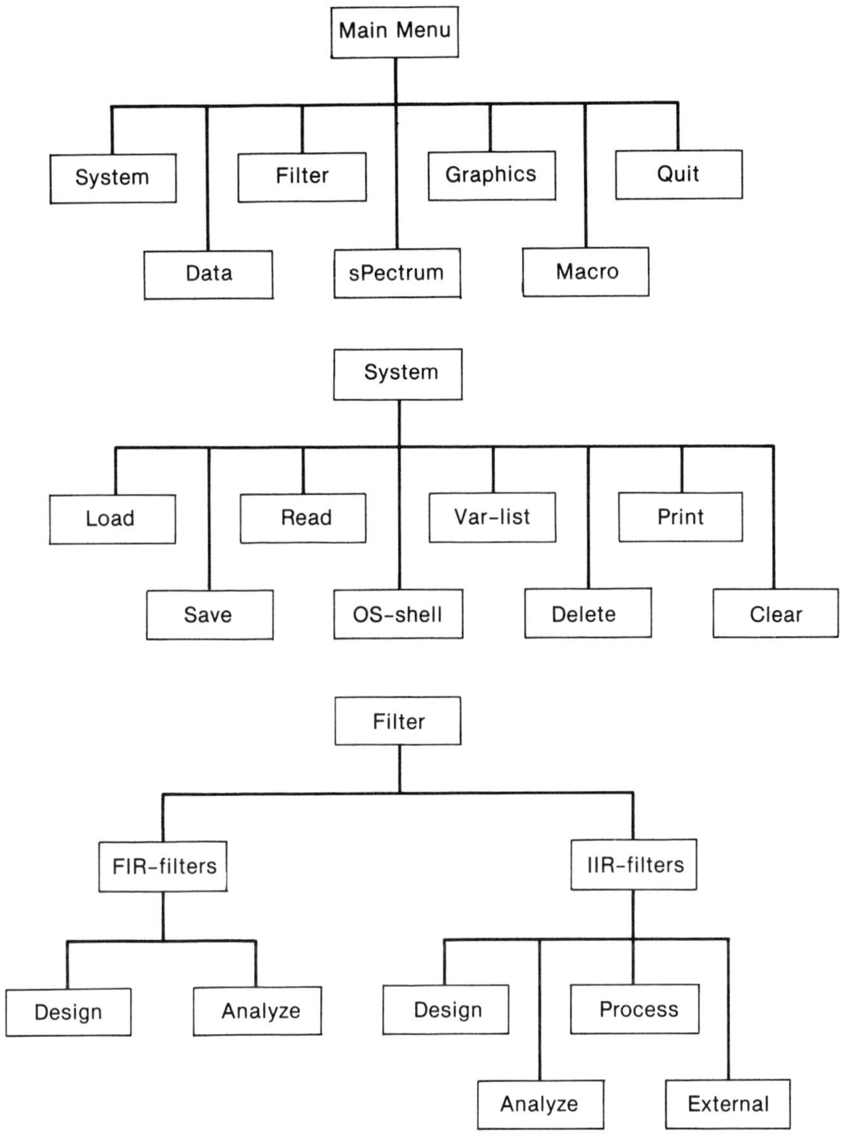

Figure 2-3a PC-DSP menu tree.

keys to make selections. Later on, when you get a better feel for the menu system, you may want to switch to the key-letter method, since it is quite faster.

Context-sensitive on-line help can be obtained at any point in the program by simply pressing the [F1] function key. A help window is "pasted" on top of the work screen. When you are done reading the help window, you can press the <ESC> key to

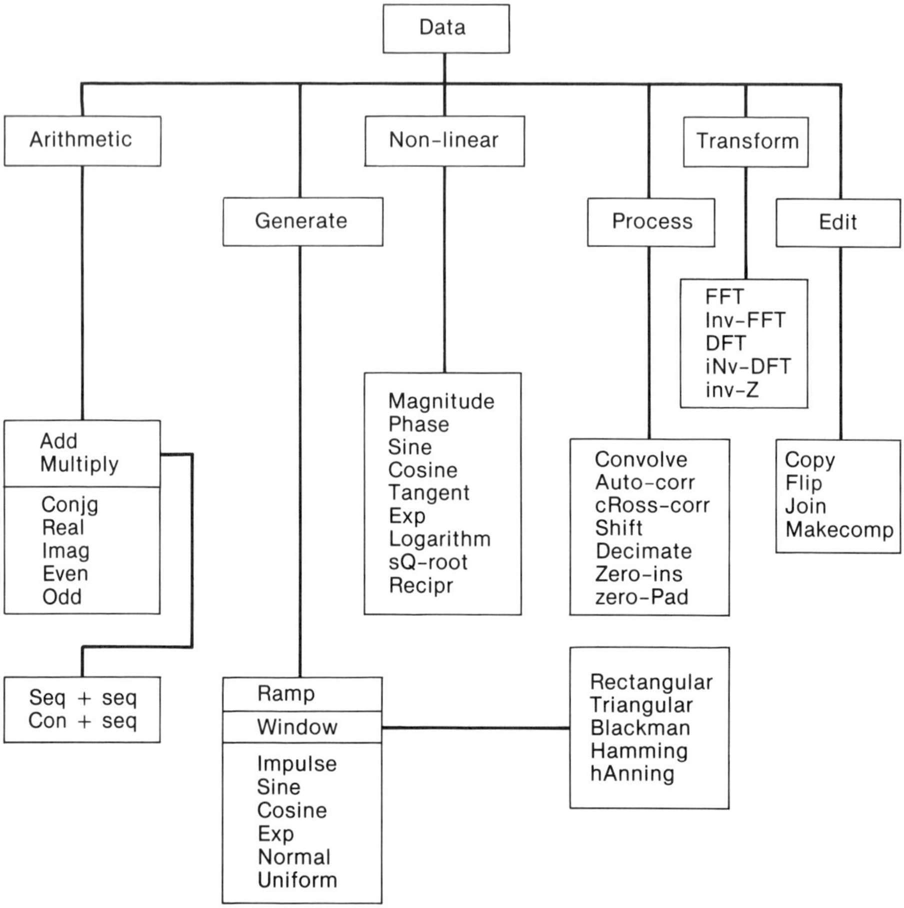

Figure 2-3b PC-DSP menu tree.

remove it and continue from where you left off. For some subjects, the help information provided is longer than one window page. In such a case, up and down arrow keys can be used to scroll through the help file and view it one page at a time. Pressing the [F1] function key again while the help window is on the screen provides general help information about PC-DSP that is several pages long.

2-2 PC-DSP VARIABLES

In PC-DSP, the basic data variable is a "sequence" that consists of a number of samples that are either real- or complex-valued. PC-DSP variables are identified with alphanu-

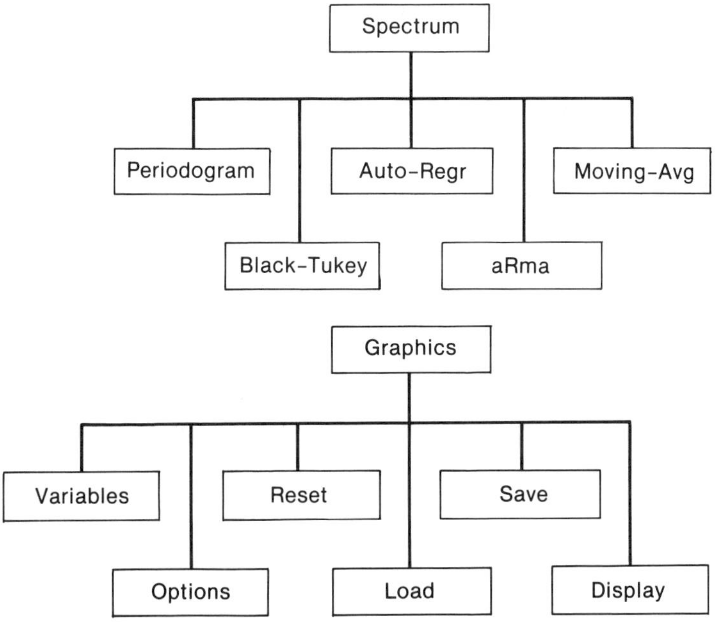

Figure 2-3c PC-DSP menu tree.

meric names that may be up to eight characters long. The first character in the name of a sequence is required to be nonnumeric. Also, variable names may not contain any punctuation characters. Most PC-DSP operators operate on sequences, and they also generate sequences as their results. As an example, the notation

```
Var1 = FFT { Var2 }
```

indicates that the FFT (Fast Fourier Transform) operator is to be applied to the variable 2, and the resulting transform sequence is to be stored in the variable 1. The user will be asked to fill in the names of the two variables involved.

Any particular sequence can be up to 4096 samples long. PC-DSP keeps track of up to 50 named sequences or up to a total of 16,000 real numbers. Each sequence that is either generated within the program as a result of an operation or loaded from an external disk file is kept in memory until it is deleted by the user, overwritten by another sequence that has the same name, or until the program is terminated.

We are now ready to illustrate the variable-handling capabilities of PC-DSP. Let's begin by choosing the *System* component from the main menu. This brings up the *System* menu in the upper left part of the screen (see Fig. 2-4). Use of the *System* menu is similar to the main menu, except for the fact that cursor movement is in the vertical direction.

The first function in the *System* component is *Load*, which is used for loading a sequence that was previously saved into a disk file. We will use this function to retrieve the sequence named RESPONSE, which is in the data file MYDATA1.DSP. Now move the highlight to the *Load* function, and press the <ENTER> key.

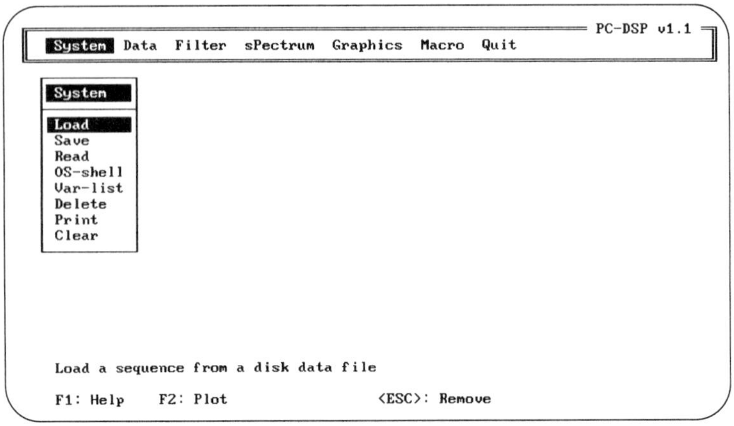

Figure 2-4 System menu.

When *Load* is selected, a data entry window is displayed in the right half of the screen. The program first prompts you for the type of the file to be loaded.

```
File type             :  D
                        ----
```

A little box to the right of the prompt line displays the two possible choices:

```
D  =  DSP file (binary)
A  =  ASCII file (text)
```

The default file type it is a DSP file, and it is indicated by the letter *D* in the user response area. Whenever PC-DSP requires a one-letter answer, possible choices are listed in a box similar to the one shown, and a default answer is suggested. Pressing the <ENTER> key without entering any characters indicates acceptance of the default answer. PC-DSP data file types will be covered in detail in Chapter 3. For the time being, it will suffice to say that the file we are about to load is a DSP file. So, just press the <ENTER> key to accept the default file type.

The next prompt of the program is

```
Filename [.DSP] : ______________
```

Pay attention to the status line. Before any characters are typed in response to the file name prompt, the middle part of the status line reads

```
<ESC>:  Menu
```

At this point, if you change your mind and decide not to go on with the *Load* function, you can press the <ESC> key to remove the data entry window and return to the *System* menu.

Now, type the name of the file to be loaded—i.e., MYDATA1—but do not press the <ENTER> key yet. Watch the status line. The definition of the <ESC> key changes to

```
<ESC>:  Re-enter
```

Pressing the <ESC> key now would cause the user response MYDATA1 to be wiped out so that it could be reentered. This is useful in case of an erroneous entry. Of course, the backspace key can also be used for the same purpose. It doesn't wipe out the entire response, however, it just deletes the last character typed.

If you have correctly typed the file name, press the <ENTER> key now. It is not necessary to type the three-letter extension for the file name, since the program uses the default extension DSP. After the disk drive spins for a few seconds, you should get the following message:

```
Variable = RESPONSE , Type = R , Size = 101
```

This is PC-DSP's way of indicating that it just added a sequence named RESPONSE to the contents of its memory. This sequence is real-valued and contains 101 samples.

Again, the middle part of the status line reads

```
<ESC>:  Menu
```

The *Load* operation has been completed. Now, we need to press the <ESC> key to remove the data entry window and return control to the *System* menu. After the <ESC> key is pressed, the status line changes to

```
<ESC>:  Remove
```

When this message is displayed, pressing the <ESC> key removes the current menu (in this case, the *System* menu) and returns control to the menu that is one level higher (in this case, the main menu).

As a convenience feature, the space bar can be used in place of the <ESC> key. (Some people find it easier to reach.) The only exception is during data entry, when the <ESC> key must be used.

To get a quick glimpse of the sequence that's just been loaded, press the [F2] function key. Assuming that your system has graphics capability, your screen should look similar to Fig. 2-6. The sequence is graphed in discrete mode, using default settings for axis limits and data labels. We will call this a quick-plot of the sequence RESPONSE. The [F2] function key generates a quick-plot of the most recently accessed plottable sequence. (A plottable sequence is one that is real-valued and shorter than 1025 samples.) In Chapter 5, alternative methods will be discussed for producing graphs with customized graph settings.

Now press <ESC> to remove the quick-plot from the screen and return control to the *System* menu. Select the *Var-list* function. The program should produce a list of variables stored in memory, as shown in Fig. 2-7.

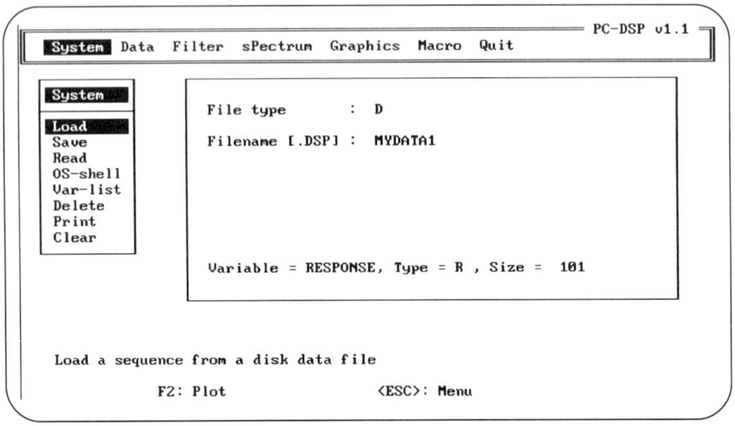

Figure 2-5 Loading the sequence RESPONSE.

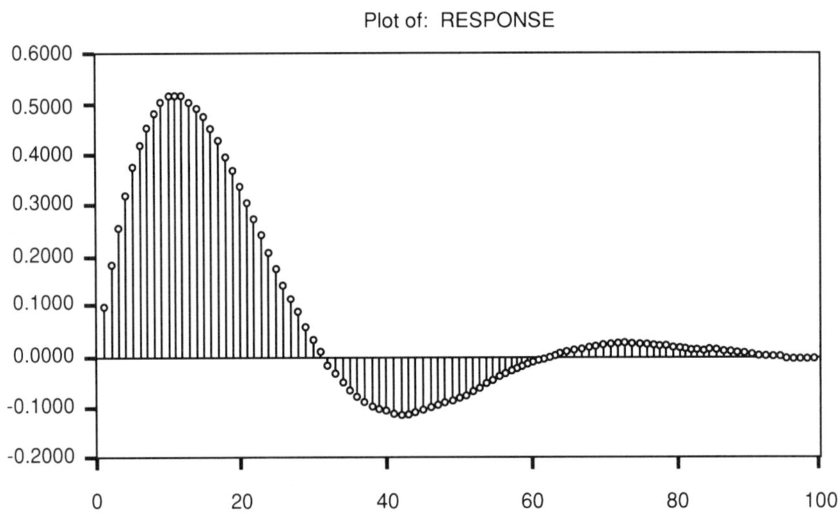

Figure 2-6 Quick-plot of the sequence RESPONSE.

At this point, the only variable in memory is the one that has just been loaded (i.e., RESPONSE). If we had other variables stored in memory, they would also show up in this list.

Press the <ESC> key to go to the *System* menu again, and select the *Print* function. This function is used for printing sample values of a sequence either on the screen or on the printer. After *Print* is selected, the first prompt of the program is

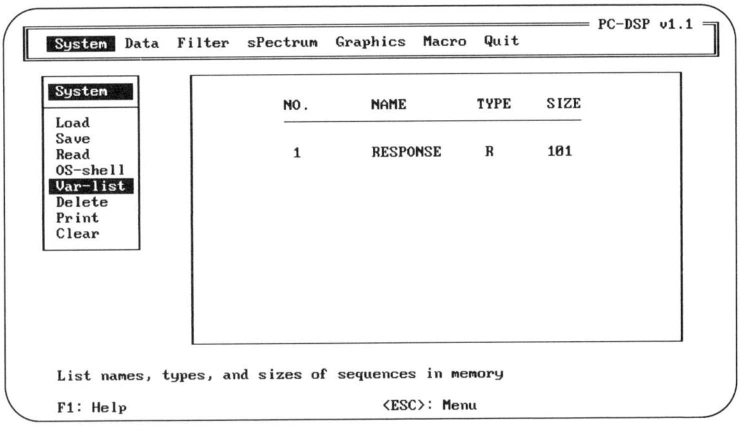

Figure 2-7 Listing variables in PC-DSP.

```
Variable name          :          RESPONSE
                                  ----------
```

where the default answer displayed is RESPONSE. In general, the name of the most recently accessed variable is used as the default. Press the <ENTER> key to accept the default answer.

Next, the destination needs to be specified.

```
Print to        :  S         S = Screen
                             P = Printer
```

Just press the <ENTER> key to accept the default answer S for a screen print that results in the display shown in Fig. 2-8. Note that only the first 10 samples of the sequence RESPONSE are displayed in tabular form. We will consider this to be one screen page. PgUp and PgDn keys on the numeric keypad can be used to scroll through pages of 10 samples each and see the rest of the sequence. In addition, up and down cursor keys can be used to scroll the display one sample at a time. Also note that RESPONSE is tabulated as a complex sequence, although it is real-valued. In PC-DSP, all sequences are treated as complex, and a real sequence is taken as a special case of a complex sequence where the imaginary part is equal to zero.

If you have a printer connected to the computer, you may want to obtain a hard copy of the sequence RESPONSE by answering the prompt

```
Print to        :  P
```

In this case, the program will ask for the range of samples to be sent to the printer. Just accept the default values to print the entire sequence, or enter a range to print part of it.

PC-DSP allocates random access memory for storage of up to 16,000 real samples. It also allocates memory for storing name, type, and size information of a maximum of

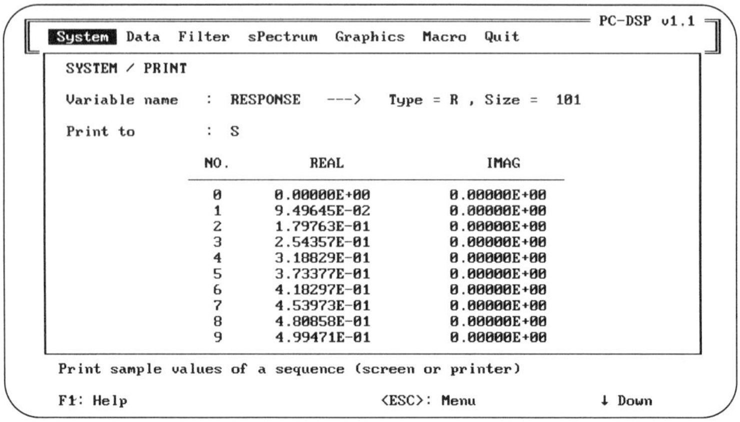

Figure 2-8 Screen print of the sequence RESPONSE.

50 variables. For most types of problems encountered in a typical senior-level digital signal processing course, this is probably more than you will need. Nevertheless, if you ever see one of the error messages

```
ERROR: Memory full .  Press <ESC>
```

or

```
ERROR: Too many variables .  Press <ESC>
```

displayed on the status line, then it is necessary to make space by deleting some variables that are no longer needed. This can be done using the *Delete* function of the *System* menu. When this function is selected, the program will ask for the name of the variable to be deleted.

```
Variable name          :       RESPONSE
                              ----------
```

PC-DSP variable names are insensitive to the case. The entries RESPONSE, response, Response, and RESponse all refer to the same variable. As soon as a variable name is entered, PC-DSP searches its data memory for that variable, and if it is found, its type and size are displayed on the next line, along with a request for confirmation.

```
Type = R , Size = 101
Do you want to delete ?  (Y/N) :  N
                                 ---
```

As a safety measure against accidental deletion of a variable, the default answer to the last question is *N* for no. To delete the variable RESPONSE, you will need to override this default answer with a *Y* for yes. If you decide not to delete RESPONSE, however, you can just accept the default answer N by pressing the <ENTER> key. The *Delete* function affects only variables in memory. It does not delete a variable saved into a disk file.

The *Clear* function of the *System* menu can be used to delete all the variables in memory and restore PC-DSP to its initial state. This function is useful when you need to clear the memory before starting to solve a new problem.

Finally, we briefly mention the *OS-shell* function of the *System* component. This function is used for temporarily returning control to the disk operating system (DOS) without losing the data in PC-DSP. When the *OS-shell* function is selected, PC-DSP displays the message

```
Type EXIT to return to PC-DSP
```

and leaves control to DOS. At this point, DOS commands can be issued to get a directory listing, to copy files, or to run another program. When you are ready to resume with PC-DSP, type EXIT at the DOS prompt.

The *OS-shell* function should be used with care, especially when running large application programs. During a *OS-shell*, PC-DSP stays in memory and occupies about 350 k-bytes for program and data storage. If you load an application program that does not fit in the remaining memory space above PC-DSP it may overwrite part of PC-DSP, and cause the system to hang up when you attempt to return to PC-DSP. Also, in order to return from a *OS-shell* successfully, the processor must be able to locate the COMMAND.COM file of the disk operating system.

3

Data Generation, Input and Output

In most digital signal processing problems, one needs to generate samples of a particular data sequence for use as input to a system or a transformation. This chapter concentrates on the waveform synthesis capabilities of PC-DSP, i.e., generation of user-specified discrete data sequences. We also discuss methods for saving these sequences into disk data files and later retrieving them.

3-1 SAMPLE-BY-SAMPLE ENTRY OF SEQUENCES

One way to generate a sequence in PC-DSP is to type in its sample values through the keyboard one sample at a time. Although this could be a tedious undertaking for a long sequence, it is the easiest entry method for short data sequences (typically, sequences with fewer than 20 samples) or for sequences that can not be represented by analytical expressions. Keyboard entry of a sequence is achieved by using the *Read* function of the *System* menu. Up to 60 real- or complex-valued samples can be entered at a time. We will explain the use of the *Read* function with two examples.

Example 3-1:

Generate a five-point sequence $x(n)$ with the following sample values:

$x(n) = \{3\ 2\ \ -2\ 0\ 7\ \}, \qquad n = 0, ..., 4$

Select the *Read* function from the *System* menu and answer the program prompts as follows:

```
Variable name    :   XN
Sample values    :   3; 2; -2; 0; 7
```

Note that entered samples are separated from each other by semi-colons. PC-DSP should respond with the familiar message

```
Variable = XN        , Type = R , Size =     5
```

You can now do a Var-list or Print to verify that the sequence XN is indeed stored in memory.

Example 3-2:

The *Read* function can also be used to enter complex-valued sequences through the keyboard. In this example, we will generate a complex sequence $w(n)$ with the following samples:

$$w(1) = -1.2 + j3.2 \quad w(6) = -j4.0$$

$$w(2) = 2.5 + j5.0 \quad w(7) = -1.1 + j3.5$$

$$w(3) = 4.4 \quad w(8) = 0.5 + j2.2$$

$$w(4) = 3.0 - j2.0 \quad w(9) = -6.5 + j3.2$$

$$w(5) = 1.5 + j1.5 \quad w(10) = 0.0$$

In PC-DSP, a complex number is entered with its real and imaginary parts separated with a comma. For example, $-1.2+j3.2$ is entered as $-1.2, 3.2$. Thus, a semicolon separates consecutive samples of a sequence, whereas a comma separates real and imaginary parts of a sample. If a sample is purely imaginary, then it is sufficient to enter just its imaginary part preceded by a comma. For example, the number *j5* can be entered as 0,5 or as ,5.

When using the *Read* function, if there are more samples than would fit on one line, then end that line with a semicolon, and press the <ENTER> key (see Fig. 3-1). PC-DSP will then prompt you for another line so that more samples can be entered. This can be repeated for a maximum of 6 lines, or 60 samples.

3-2 GENERATING LINEAR RAMP SEQUENCES

Linear ramp sequences are generated using the *Ramp* function in the *Generate* submenu of the *Data* component. This function generates a sequence with specified number of samples, initial value, and increment. Specifically, if the initial value and the increment are C_1 and C_2, respectively, then samples of the sequence are computed as

$$x(n) = C_1 + nC_2 \qquad \text{for } n = 0, ..., N-1$$

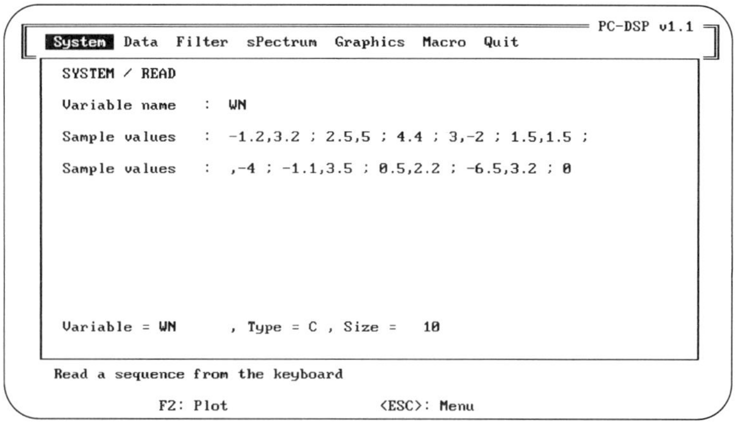

Figure 3-1. Keyboard entry of a complex-valued sequence.

In general, both the initial value and the increment may be complex-valued parameters.

Example 3-3:

Generate a real-valued linear ramp sequence RMP with amplitudes going from 0 to 10 in 100 samples.

The problem statement does not specify the increment, but the final value (i.e., 10) is given instead. There are 99 sampling intervals in 100 samples; therefore, the increment must be 10/99. Now select the *Data* menu, then the *Generate* submenu, and finally the *Ramp* function. At the top of the data entry window, the format of this function is displayed.

```
Varl = Init + n.Incr   ,  n = 0, ..., #Sampl-1
```

This format statement indicates that the *Ramp* function accepts three arguments, namely, Init, Incr, and #Sampl. It generates a sequence in return. Answer the prompts as shown below:

```
Varl = Init + n.Incr   ,  n = 0, ..., #Sampl-1
Varl   :  RMP
Init   :  0
Incr   :  10/99
#Sampl:   100
```

This results in the familiar response

```
Variable = RMP      , Type = R , Size =  100
```

to be displayed in the lower part of the data entry window. A quick-plot of the generated sequence can be obtained by pressing the [F2] function key (Fig. 3-2). Pay special attention to the way the increment is entered. Instead of calculating the value 10/99 and entering it as 0.10101, we let the program evaluate the arithmetic formula. PC-DSP accepts arithmetic expressions on the answer line for numerical entries. All the following are valid entries when a program prompt calls for a floating-point value to be entered.

```
3.141593/256        (= 0.012272)
2*3.14/25+2         (= 2.2512)
3*5+2,-1/10         (= 17 - j0.1)
```

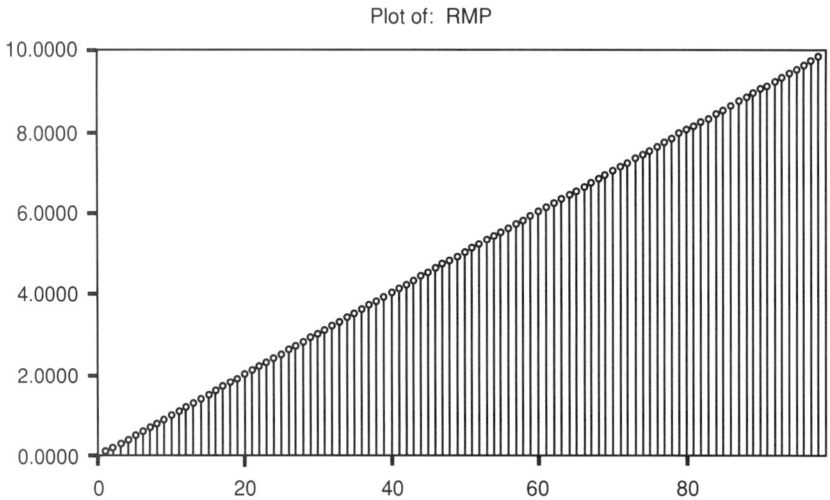

Figure 3-2. Linear ramp sequence RMP of Ex. 3-3.

3-3 EDITING DISCRETE SEQUENCES

In the *Data* component of PC-DSP, several functions are available for "cut-and-paste" type editing of discrete sequences. These functions are *Copy, Flip, Join,* and *Makecomp*, and they are located in the *Edit* submenu of the *Data* component. They are mostly used for step-by-step synthesis of complicated discrete waveforms. In this section, we discuss the use of each one of these functions through example problems.

The *Copy* function generates a variable by duplicating samples of another variable within a specified range. For example, if we have a variable $x_2(n)$ in memory with sample values

$$x_2(n) = \{3 \;\; -2 \; 4 \; 0 \;\; -7 \; 2 \; 5\}, \qquad n = 0, ..., 6$$

copying samples 2 through 5 would generate a new variable, say $x_1(n)$, with samples

$$x_1(n) = \{4 \; 0 \;\; -7 \; 2\}, \qquad n = 0, ..., 3$$

It is important to note that the input sequence $x_2(n)$ remains unchanged after the *Copy* operation, i.e., the four samples of $x_1(n)$ are not actually "taken away" from $x_2(n)$ but are just copied into $x_1(n)$.

Example 3-4:

Generate a 15-point linear ramp sequence with an initial value of 0.0 and an increment of 10/99.

Since we have already generated a linear ramp sequence with an increment of 10/99 in Ex. 3-3, we will use that sequence and just copy the number of samples needed. Select the *Copy* function from the *Edit* submenu. The format of this function is

```
FORMAT:   Var1 = COPY { Var2, From, To }
```

Answer the program prompts as follows:

```
Var1 :   SEG1
Var2 :   RMP
From :   0
To   :   14
```

With these entries, samples 0 to 14 of the sequence RMP are copied into a new sequence SEG1.

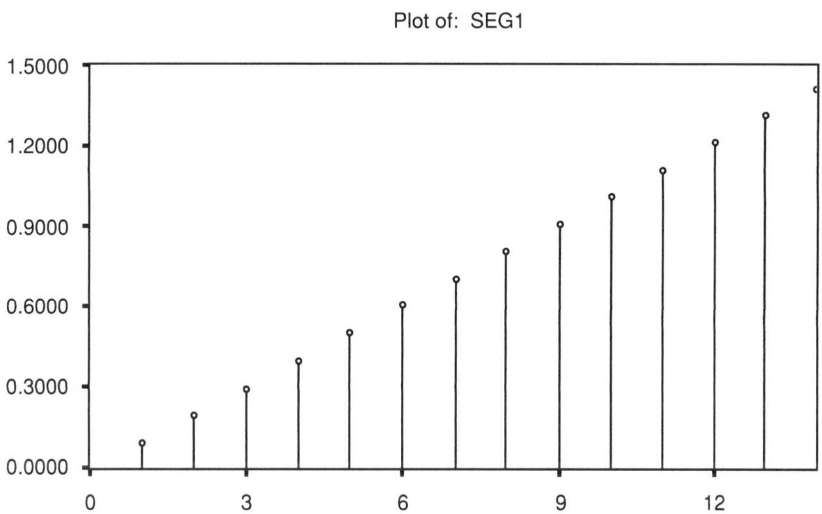

Figure 3-3. Quick-plot of SEG1 generated in Ex. 3-4.

The *Copy* function can also be used to copy part of a sequence in reverse order. The next example illustrates this point.

Example 3-5:

We will now create a new variable SEG2 by taking the middle 13 samples of SEG1 of Ex. 3-4 and writing them into SEG2 in reverse order. All we need to do is copy from 13 to 1 instead of the other way around. Select *Copy* again, and enter the following:

```
FORMAT:   Var1 = COPY { Var2, From, To }
Var1 :   SEG2
Var2 :   SEG1
```

```
From :   13
To   :   1
```

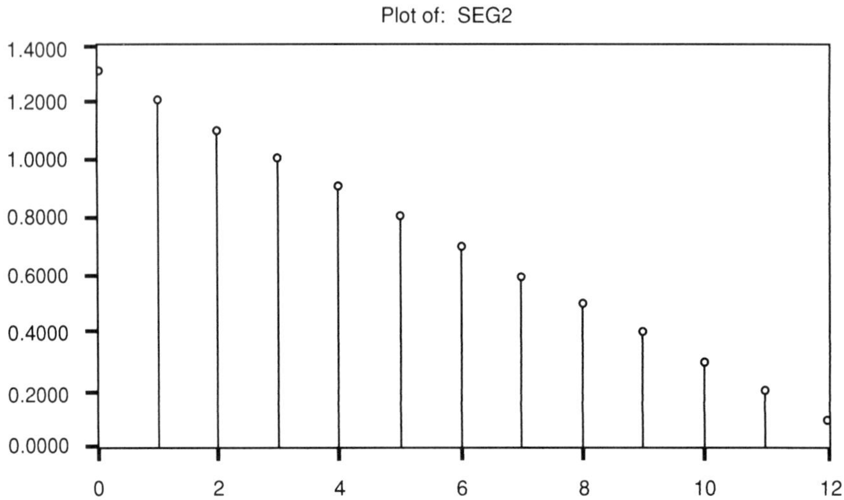

Figure 3-4. Quick-plot of SEG 2 generated in Ex. 3-5.

The *Flip* function serves a purpose similar to backward copying. It generates a sequence by duplicating samples of another sequence in reverse order. The main difference between *Flip* and *Copy* is that *Flip* operates on the entire sequence, whereas *Copy* can be used on a specified range of samples.

Example 3-6:

We will now generate a sequence RMP2 by reversing the order of samples in the sequence RMP obtained in Ex. 3-3. Select the *Flip* function. The format is displayed at the top of the data entry window.

```
Var1 = FLIP { Var2 }
```

Enter the names of the variables involved as follows:

```
Var1  :   RMP2
Var2  :   RMP
```

A quick-plot of the sequence RMP2 is shown in Fig. 3-5, which is a mirror image of RMP.

The *Join* function is used for generating a sequence made of two sequences, placed one after the other. The format for this function is

```
Var1 = JOIN { Var2, Var3 }
```

The sequence Var1 is constructed by first taking all samples of Var2 in their normal order, and appending to them the samples of Var3, again in their normal order. Thus,

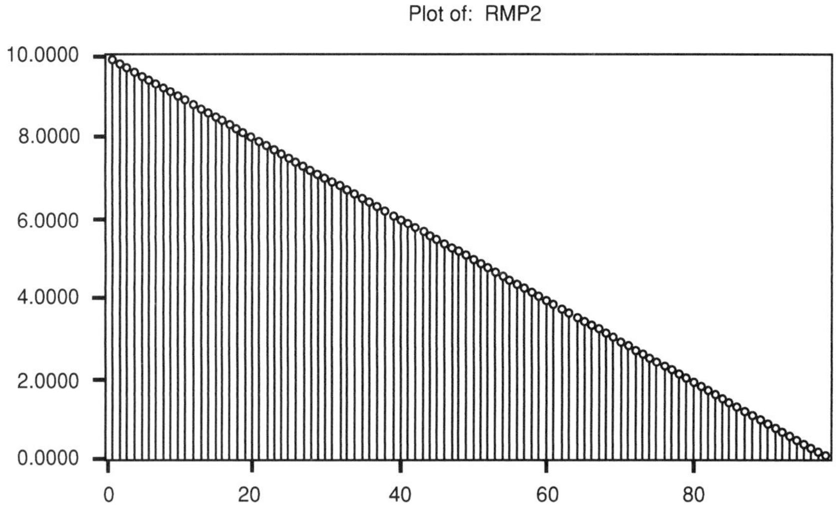

Figure 3-5. Reversed ramp sequence RMP2 of Ex. 3-6.

the length of the new sequence Var1 is the sum of lengths of Var2 and Var3. Before using *Join*, it should be ensured that this total length does not exceed 4096, since this is the maximum number of samples that can be stored in one sequence.

Example 3-7:

The results of Ex. 3-4 and Ex. 3-5 can be used to generate one period of a sawtooth waveform. We will simply *Join* the two segments SEG1 and SEG2 to construct the waveform WAVE. Select the *Join* function, and fill in the variable names, as shown:

```
FORMAT:  Var1 = JOIN { Var2, Var3 }
Var1 :    WAVE
Var2 :    SEG1
Var3 :    SEG2
```

The two joined sequences, SEG1 and SEG2, had 15 and 13 samples, respectively. The sequence WAVE is 28 samples long and is shown in Fig. 3-6.

Example 3-8:

A sequence may also be joined with itself to generate a second period. Let's select *Join* again, and do the following:

```
FORMAT:  Var1 = JOIN { Var2, Var3 }
Var1 :    WAVE
Var2 :    WAVE
Var3 :    WAVE
```

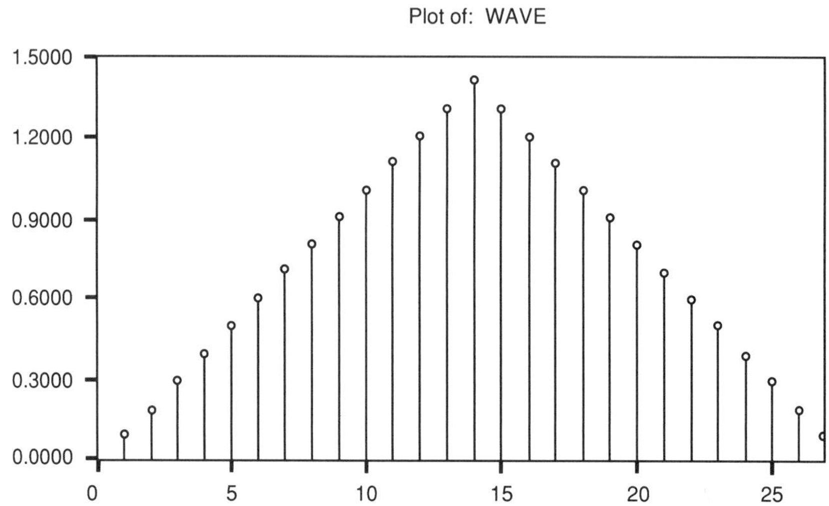

Figure 3-6. Quick-plot of WAVE generated in Ex. 3-7.

In the preceding development, the 28-point sequence stored under the name WAVE is joined with itself, and the resulting 56-point sequence is stored under the name WAVE, replacing the original 28-point sequence, as shown in Fig. 3-7. The idea is very similar to an arithmetic expression such as

```
A = A + A
```

in FORTRAN or BASIC programming languages, where the arithmetic operation on the right side is performed and the result is written into the variable on the left side.

The *Makecomp* function is used for combining two real sequences into one complex sequence. Specifically, if $x_2(n)$ and $x_3(n)$ are two real sequences, then $x_1(n)$ is computed as

$$x_1(n) = x_2(n) + jx_3(n)$$

If the two input sequences that make up the real and imaginary parts are not of the same length, then the shorter sequence is padded with zeros to make the two inputs compatible.

Example 3-9:

Generate 20 samples of the complex sequence

$$\delta(n) + ju(n) \qquad \text{for } n = 0, ..., 19$$

We will first generate the impulse sequence $\delta(n)$ for the real part. Then, we will generate the unit-step sequence $u(n)$ for the imaginary part, and finally combine the two using the *Makecomp* function.

The *Generate* submenu of the *Data* component contains an *Impulse* function that can be used to generate a specified number of samples of a unit-impulse sequence. Select *Impulse*, and make the following entries:

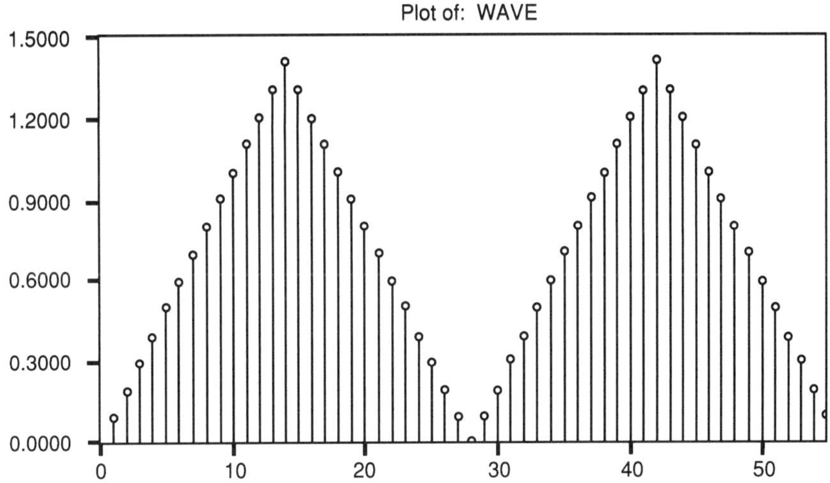

Figure 3-7. Quick-plot of WAVE generated in Ex. 3-8.

```
FORMAT:   Var1 = IMPULSE { #Sampl }
Var1   :   XREAL
#Sampl:    20
```

The impulse sequence is now stored in the variable XREAL. To generate the unit-step sequence for the imaginary part, the *Ramp* function can be used with unit initial value and zero increment. Select *Ramp,* and make the following entries:

```
Var1 = Init + n.Incr  ,  n = 0, ..., #Sampl-1
Var1    :   XIMAG
Init    :   1
Incr    :   0
#Sampl  :   20
```

At this point, the imaginary part is in the variable XIMAG. We are ready to use the *Make-comp* function to combine XREAL and XIMAG into a complex sequence XCOMP.

```
FORMAT:  Var1 = Var2 + j Var3
Var1 :   XCOMP
Var2 :   XREAL
Var3 :   XIMAG
```

Since the resulting sequence XCOMP is complex, it cannot be plotted using the [F2] function key. Nevertheless, you can print it using the *Print* function of the *System* component, as explained in Sec. 2-2.

3-4 TRIGONOMETRIC AND EXPONENTIAL FUNCTIONS

The *Generate* submenu of the *Data* component contains functions that generate trigonometric or exponential sequences that are often needed in digital signal processing problems. These functions are *Sine, Cosine,* and *Exp*. All three have the identical format,

```
Var1 = A SIN { B n + C } ,   n = 0, ..., #Sampl-1
```

where COS or EXP may also be substituted for SIN. The three constants A, B, and C may be real- or complex-valued. As usual, the number of samples that can be generated is limited to 4096.

Example 3-10:

Generate 100 samples of the signal

$$2.2\sin\left[2\pi(0.1)n + \frac{\pi}{4}\right]$$

for $n = 0, ..., 99$.

Looking at the format of the sine function, it is easy to see that the desired sequence can be generated by choosing scalar parameter values $A = 2.2$, $B = 0.2\pi$, and $C = \pi/4$. Select the *Sine* function, and make the following entries.

```
Var1 = A SIN { B n + C } ,   n = 0, ..., #Sampl-1
Var1   :  TONE
   A   :  2.2
   B   :  0.2 * π
   C   :  π/4
#Sampl :  100
```

Note that the character π can be entered using the key combination Alt-P, i.e., by holding the Alt key down while typing the letter *P*. The quick-plot is shown in Fig. 3-8.

3-5 RANDOM SEQUENCES

Random phenomena play an important role in digital signal processing problems. A large number of systems encountered in the applications of digital signal processing have stochastic properties. Random noise is part of virtually every signal. In computer simulation of discrete-time systems, we usually need to simulate these random effects as well.

The *Generate* submenu of the *Data* component of PC-DSP contains two random sequence generators. One is the uniform random number generator *Uniform*, and the other is the normal (Gaussian) random number generator *Normal*. It should be stressed however that these functions generate *pseudo-random* sequences. Obviously, everything that goes on within the microprocessor of a computer is deterministic (well, almost), so these random sequences are generated by a deterministic set of software routines. They just simulate random signals within specified statistical constraints.

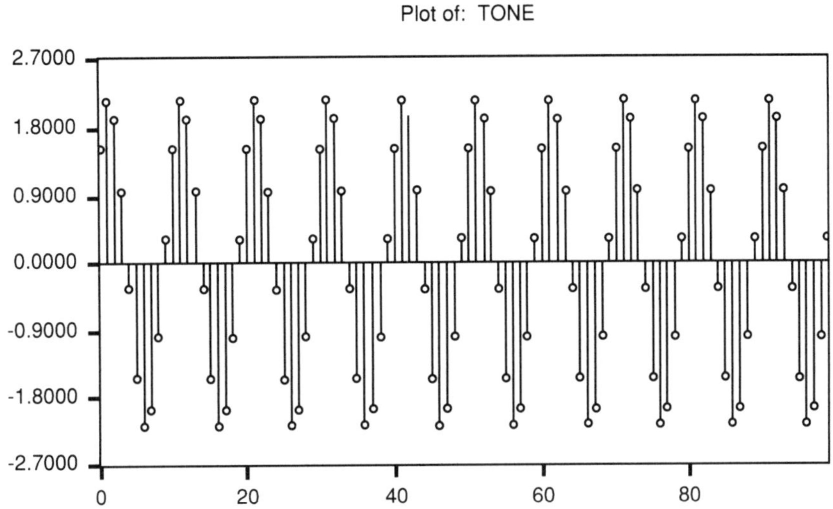

Figure 3-8 The sine waveform generated in Ex. 3-10.

A uniform random signal is characterized by a probability density function (pdf) of the form

$$p(x) = \begin{cases} \dfrac{1}{b-a}, & \text{if } a < x < b \\ 0 & \text{otherwise} \end{cases}$$

and the pdf of an ideal normal (Gaussian) process is

$$p(x) = \frac{1}{\sqrt{2\pi}\sigma} \exp\left[\frac{(x-m_x)^2}{2\sigma^2}\right]$$

Computational details of simulating uniform and normal random distributions can be found in Appendix A. We present two examples of generating random sequences.

Example 3.11:

Generate a sequence of 56 random numbers uniformly distributed between -0.5 and 0.5. Then add this sequence to the sawtooth waveform of Ex. 3-8 to obtain a noisy sawtooth waveform.

Begin by selecting the *Uniform* function. Enter the following parameter values to generate the uniform random sequence NOISE1.

```
FORMAT:  Var1 = UNIFORM { Lower, Upper, #Sampl }
Var1   :   NOISE1
Lower  :   −0.5
Upper  :   0.5
#Sampl:    56
```

The program should respond with the familiar message

```
Variable = NOISE1    , Type = R , Size = 56
```

Pressing the [F2] key produces the quick plot in Fig. 3-9.

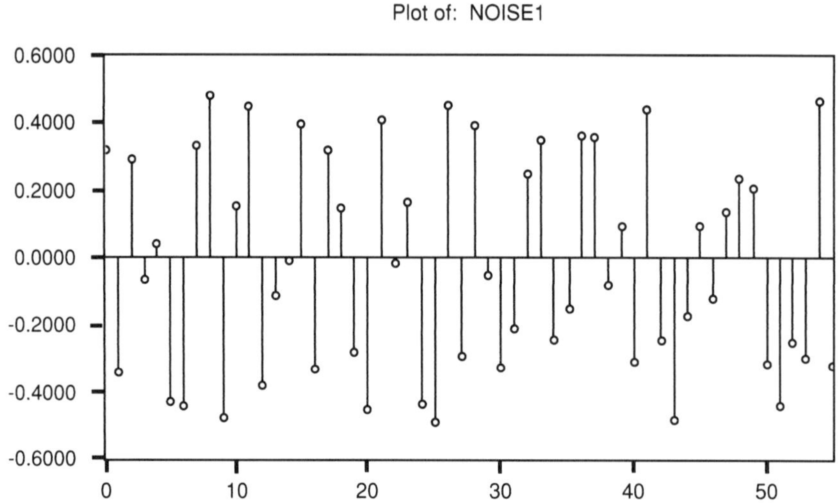

Figure 3-9 Uniform random sequence generated in Ex. 3-11.

We will use the *Add* function of the *Arithmetic* submenu to add this random sequence to the sawtooth waveform of Ex. 3-8. Select *Arithmetic*, and then select *Add*. This brings up another submenu with two items: *Seq+seq* and *Con+seq*. The first one is for adding two sequences sample by sample, and the second is for adding a specified constant to each sample in a sequence. Since we wish to add two 56-point sequences, select *Seq+seq*. The format is

```
FORMAT:  Var1 = Var2 + Var3
```

Make the following entries:

```
Var1  :  NWAVE
Var2  :  WAVE
Var3  :  NOISE1
```

The resulting noisy sawtooth waveform is shown in Fig. 3-10.

Example 3-12:

Generate a sequence of 100 samples from a normal distribution with zero mean and unit variance.

Begin by selecting the *Normal* function of the *Generate* submenu. The parameters that need to be specified are the mean, the variance, and the number of samples. The values of these three parameters for this particular example are 0, 1, and 100, respectively, and all three happen to be the defaults of the program. So enter the variable name, and press

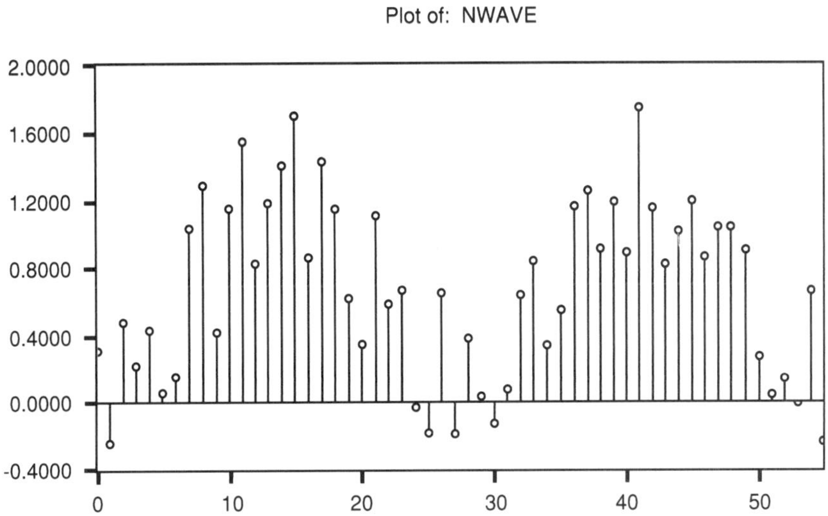

Figure 3-10 Noisy sawtooth waveform obtained in Ex. 3-11.

<ENTER> three times to accept the defaults to generate the sequence NOISE2 shown in Fig. 3-11.

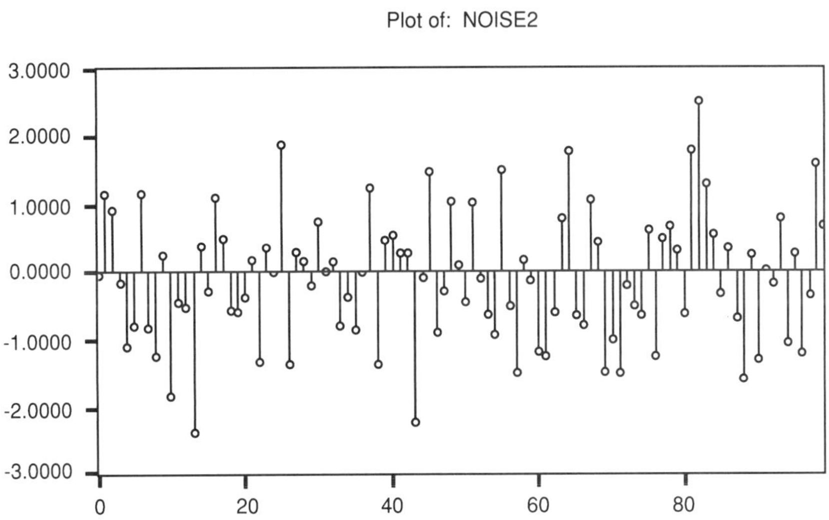

Figure 3-11. Normal random sequence generated in Ex. 3-12.

```
Var1 = NORMAL { Mean, Variance, #Sampl }
Var1      :   NOISE2
Mean      :   0
Variance  :   1
#Sampl    :   100
```

3-6 DISCRETE WINDOW FUNCTIONS

Five popular window sequences, Rectangular, Triangular, Blackman, Hamming, and Hanning windows, are built into the applicable filter design and spectrum analysis routines of PC-DSP. They are also provided in the *Generate / Window* sub-menu of the *Data* component, so they can be studied and analyzed separately if desired. We leave the computational details of window functions to the Appendix A and present an example of generating a window sequence.

Example 3-13:

Generate a Blackman window with 50 samples. First select *Generate*, then *Window*, and then *Blackman*.

```
FORMAT:  Var1 = BLACKMAN { #Sampl }
Var1   :   BLWIN
#Sampl:    50
```

We now have a 50-point Blackman window stored under the name BLWIN. A quick-plot is shown in Fig. 3-12.

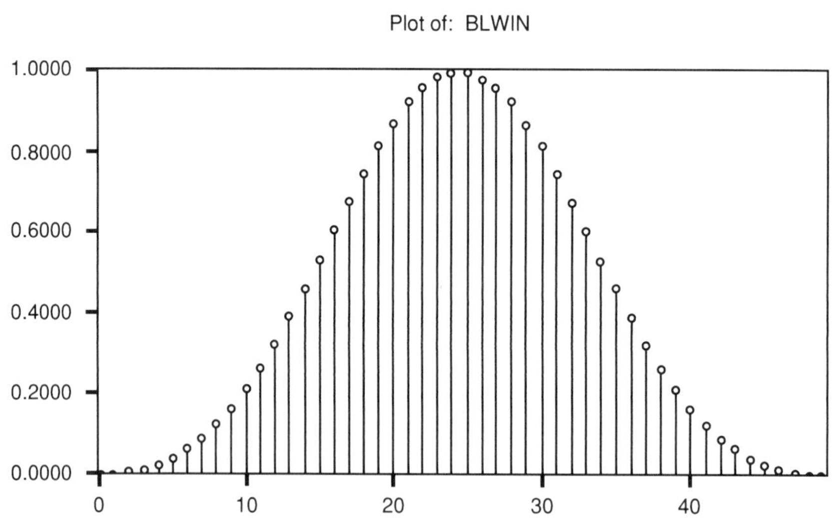

Figure 3-12. Blackman window of Ex. 3-13.

It is possible to compute the frequency spectrum of this window sequence using the Analyze function of the *FIR-filters* submenu in the *Filter* component. Details will be deferred to Chapter 6.

3-7 SYNTHESIS OF MORE COMPLICATED WAVEFORMS

In this section, we present examples of using functions of the *Data* menu to synthesize several interesting waveforms that might be encountered in signal processing problems.

Example 3-14:

Generate the sequence

$$w(n) = 3\ \sin(0.4\pi n) + 2\ \cos(0.72\pi n) \qquad \text{for } n = 0, 1, ..., 99$$

Let's first use the *Sine* function of *Generate* submenu to generate the first term with the following entries:

```
Var1 = A SIN { B n + C } ,  n = 0, ..., #Sampl-1
Var1   :  W1N
   A   :  3
   B   :  0.4 * π
   C   :  0
#Sampl:   100
```

The second term of *w(n)* can be generated in a similar way, using the *Cosine* function.

```
Var1 = A COS { B n + C } ,  n = 0, ..., #Sampl-1
Var1   :  W2N
   A   :  2
   B   :  0.72 * π
   C   :  0
#Sampl:   100
```

W1N and W2N can now be added to obtain WN. Select the *Add* function from the *Arithmetic* menu. Then, select *Seq+seq* for sample-by-sample addition.

```
FORMAT:  Var1 = Var2 + Var3
Var1   :   WN
Var2   :   W1N
Var3   :   W2N
```

The resulting 100-point sequence WN is shown in Fig. 3-13.

Example 3-15:

Generate 100 samples of the half-wave-rectified sinusoid $y(n)$ defined as

$$y(n) = \begin{cases} \sin(0.2n) & \text{if positive} \\ 0 & \text{otherwise} \end{cases}$$

This sequence can be generated in several ways. The easiest way is to first generate the full sinusoid and then add to it its absolute value. Let

$$y_1(n) = 0.5\ \sin(0.2n)$$

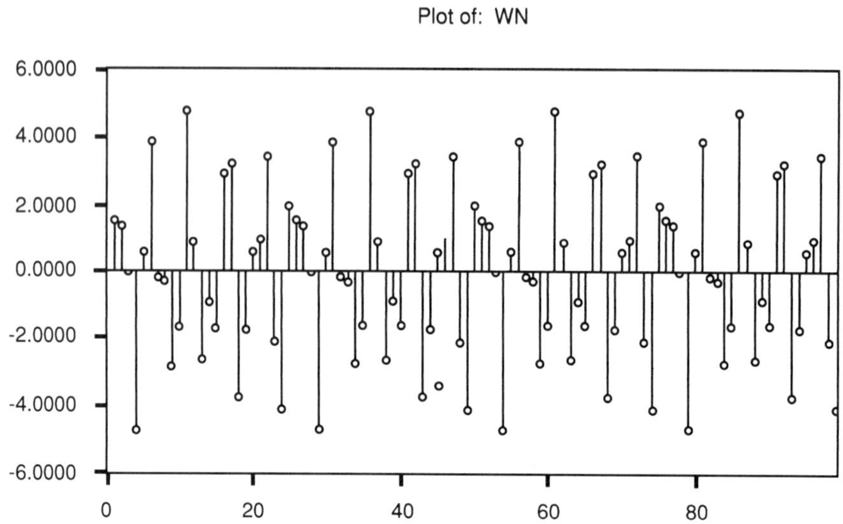

Figure 3-13. The sequence WN of Ex. 3-14.

Samples of the rectified sinusoid can be obtained as

$$y(n) = y_1(n) + |y_1(n)|$$

We will first generate the 100-point full sinusoid Y1N using the *Sine* function.

```
Var1 = A SIN { B n + C } ,   n = 0, ..., #Sampl-1
Var1   :  Y1N
   A   : 0.5
   B   : 0.2
   C   : 0
#Sampl:  100
```

To compute the absolute value sequence for Y1N, the *Magnitude* function of the *Nonlinear* submenu can be used. Select *Magnitude*, and make the following entries:

```
FORMAT:  Var1 = | Var2 |
Var1   :    Y2N
Var2   :    Y1N
```

Finally, we add the two sequences Y1N and Y2N to obtain the rectified sinusoid. Select *Arithmetic*, then select *Add*, and then *Seq+seq*.

```
FORMAT:  Var1 = Var2 + Var3
Var1   :    YN
Var2   :    Y1N
Var3   :    Y2N
```

This completes the synthesis of our rectified sinusoid. The resulting sequence WN is shown in Fig. 3-14.

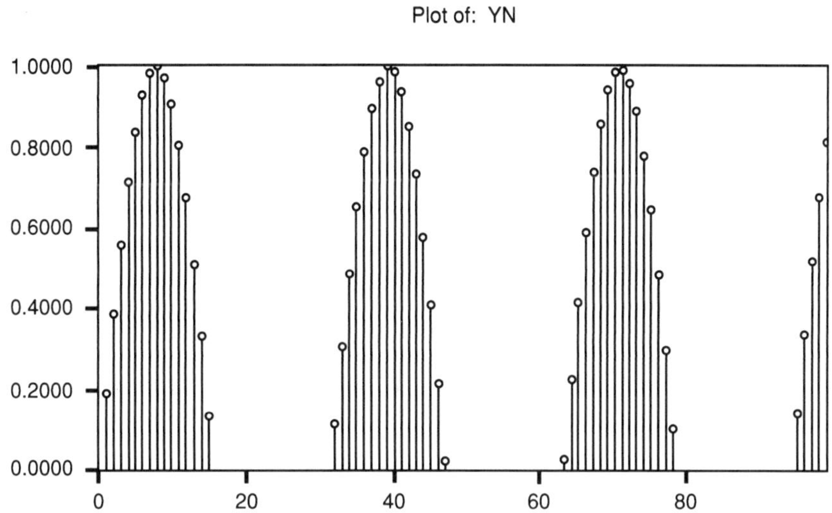

Figure 3-14. The rectified sinusoid of Ex. 3-15.

Example 3-16:

Generate the sequence

$h(n) = (-0.9)^n \qquad$ for $n = 0, ..., 99$

PC-DSP does not have a built-in function for generating a sequence in the form $h(n) = A^n$. Nevertheless, the *Exp* and *Logarithm* functions can be used for this purpose. It can be easily shown that

$h(n) = A^n = \exp[n \ln(A)] \qquad$ for $n = 0, ..., N - 1$

We will use this equation to generate the sequence $h(n)$ in five steps:

1. Generate an N-point sequence $h_1(n)$ with all samples equal to A. The easiest method is to use the *Ramp* function with initial value equal to A and increment equal to zero.
2. Compute the natural logarithm of the sequence $h_1(n)$ using the *Logarithm* function of the *Nonlinear* submenu, i.e.,

   $h_2(n) = \ln[h_1(n)]$

3. Generate $h_3(n)$ as a ramp sequence made of integer-valued samples 0, 1, ..., $N - 1$. Again, the easiest way to do this is to use the *Ramp* function.
4. Obtain $h_4(n)$ by multiplying the sequences $h_2(n)$ and $h_3(n)$ sample by sample. The *Multiply* function of the *Arithmetic* submenu can be used for this purpose.
5. Finally, compute the exponential of $h_4(n)$ to obtain the desired sequence $h(n)$. To do this, use the Exp function of the *Nonlinear* menu.

We will now apply these steps to our example. Recalling that $A = -0.9$, select the *Ramp* function and do the following.

```
Var1 = Init + n.Incr   ,   n = 0, ..., #Sampl-1
Var1   :   H1N
Init   :   -0.9
Incr   :   0
#Sampl:    100
```

For the second step, select *Nonlinear*; then select *Logarithm*.

```
FORMAT:   Var1 = LOGARITHM { Var2 }
Var1   :   H2N
Var2   :   H1N
```

Use the *Ramp* function again, this time to generate the integer sequence H3N.

```
Var1 = Init + n.Incr   ,   n = 0, ..., #Sampl-1
Var1   :   H3N
Init   :   0
Incr   :   1
#Sampl:    100
```

In the fourth step, we multiply H2N and H3N. Select *Arithmetic*, then *Multiply*, and then *Seq*seq*.

```
FORMAT:   Var1 = Var2 * Var3
Var1   :   H4N
Var2   :   H2N
Var3   :   H3N
```

For the final step, select *Nonlinear*, then select *Exp*. (Do not confuse this *Exp* function with the *Exp* function of the *Generate* submenu. The latter *generates* an exponential waveform, whereas the former computes the exponential function of a sequence.)

```
FORMAT:   Var1 = EXP { Var2 }
Var1   :   HN
Var2   :   H4N
```

The resulting sequence HN is complex, but a tabulation of sample values using the *Print* function reveals that imaginary parts are insignificant and are basically due to machine round-off errors. The real part of the sequence HN is shown in Fig. 3-15.

One interesting point in this example is the logarithm operation in step 2. We computed the natural logarithm of the sequence H1N, which had negative samples. The obvious question is, how can the logarithm of a negative number be computed?

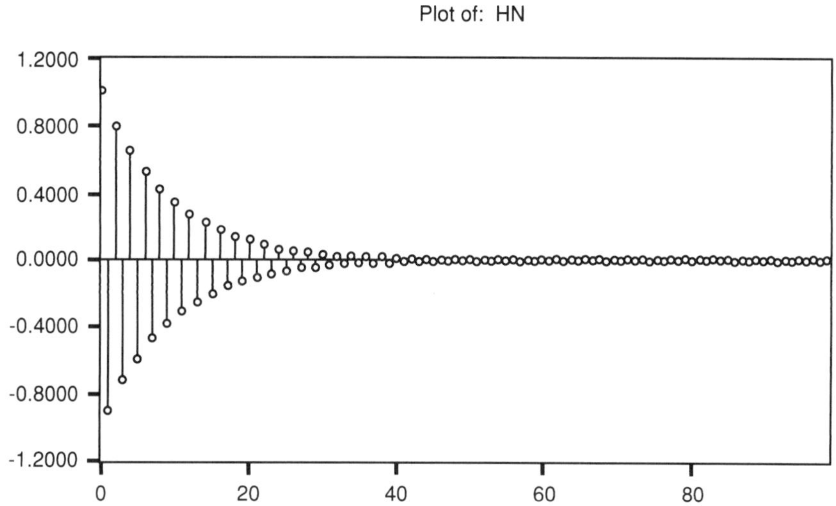

Figure 3-15. Quick-plot of the sequence HN of Ex.3-16.

The answer lies in the fact that all but a few PC-DSP functions work with complex sequences. Definitions of most traditional functions can be extended to cover complex numbers. For example, the natural logarithm of a complex number in polar form can be computed as

$$\ln(re^{j\theta}) = \ln(r) + \ln(e^{j\theta}) = \ln(r) + j\theta$$

Since any negative number can also be written as a complex number in polar form, it is possible to compute the logarithm of negative numbers if we are willing to accept complex results. (In this case, we are.) Using this philosophy, it can be shown that the natural logarithm of -0.9 is $-0.105361 + j\pi$.

Example 3-17:

The discrete message signal

$$x_m(n) = \sin(2\pi F_m n), \qquad F_m = 0.01$$

is used to modulate the amplitude of the carrier signal

$$x_c(n) = \sin(2\pi F_c n), \qquad F_c = 0.1$$

using the double sideband amplitude modulation (AM) technique with a modulation index of 0.7. The resulting signal is

$$x_{\text{AM}}(n) = [1 + 0.7x_m(n)]x_c(n)$$

Generate 256 samples of this signal for $n = 0, \ldots, 255$.

The first step is to generate the carrier signal using the *Sine* function of the *Generate* submenu.

```
Var1 = A SIN { B n + C } ,  n = 0, ..., #Sampl-1
Var1   :  XCN
   A   :  1
   B   :  0.2 * π
   C   :  0
#Sampl:   256
```

The next step is to generate the message signal using the *Sine* function again.

```
Var1 = A SIN { B n + C } ,  n = 0, ..., #Sampl-1
Var1   :  XMN
   A   :  0.7
   B   :  0.02 * π
   C   :  0
#Sampl:   256
```

Note that the modulation index 0.7 is included in the message signal, i.e., we actually have 0.7 $x_m(n)$ in the generated sequence. Now we need to add 1.0 to each sample in the sequence XMN. This can be done using the *Add* function of the *Arithmetic* submenu. Select *Arithmetic*, then *Add*, and then *Con+seq*.

```
FORMAT:  Var1 = Var2 + Const
Var1   :   XMN
Var2   :   XMN
Const  :   1
```

The sequences XCN and XMN can now be multiplied to give the amplitude modulated waveform. We will use the *Multiply* function for sample-by-sample multiplication of two sequences.

```
FORMAT:  Var1 = Var2 .  Var3
Var1   :   XAMN
Var2   :   XMN
Var3   :   XCN
```

Example 3-18:

Generate the sequence

$$p(n) = \frac{\sin(0.1n)}{0.1n} \qquad n = 0, \ldots, 99$$

Numerator and denominator sequences will be generated separately.

Let us make the following definitions:

$$q(n) = \frac{1}{0.1n} \qquad n = 1, \ldots, 99$$

$$s(n) = \sin(0.1n) \qquad n = 0, \ldots, 99$$

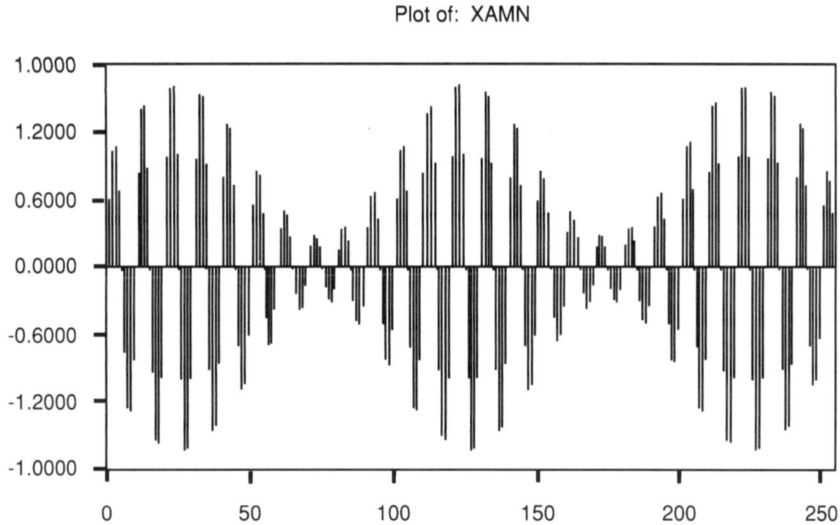

Figure 3-16 Amplitude modulated signal of Ex. 3-17.

The desired sequence $p(n)$ can be expressed as the product of $q(n)$ and $s(n)$. We already know how to generate $s(n)$ using the *Sine* function. To generate $q(n)$, we can first generate $r(n) = 0.1n$ for $n = 0, ..., 99$ using the *Ramp* function and then use the *Recipr* function of the *Nonlinear* submenu. This function computes the reciprocal of the input sequence.

There is one potential problem that needs to be addressed here. The first sample of the sequence $r(n)$ is zero, and an attempt to reciprocate $r(n)$ would result in a floating-point overflow. (You may try this if you wish. The program does not crash—it just displays an error message and refuses to do the operation that causes the overflow.) A practical solution to the overflow problem is to add a very small positive number to each sample in the sequence $r(n)$. Now, select the *Ramp* function.

```
Var1 = Init + n.Incr  ,  n = 0, ..., #Sampl-1
Var1   :   RN
Init   :   0
Incr   :   0.1
#Sampl:    100
```

Let's add 0.00001 to each sample of RN. Select *Arithmetic*, then *Add*, and then *Con+seq*.

```
FORMAT:  Var1 = Var2 + Const
Var1   :   RN
Var2   :   RN
Const  :   0.00001
```

The constant added to the sequence RN is insignificant compared with the nonzero samples of the sequence and therefore has negligible effect on our calculations. We can now proceed with the *Recipr* function.

```
FORMAT:  Var1 = 1 / Var2
Var1   :   QN
Var2   :   RN
```

Select the *Sine* function of the *Nonlinear* submenu.

```
FORMAT:  Var1 = SIN { Var2 }
Var1   :   SN
Var2   :   RN
```

Multiply QN and SN using the *Multiply / Seq+seq* function.

```
FORMAT:  Var1 = Var2 .  Var3
Var1   :   PN
Var2   :   QN
Var3   :   SN
```

The resulting sequence PN is shown in Fig. 3-17.

In step 4, the choice of the proper *Sine* function is critical. Had we used the *Sine* function of the *Generate* submenu with parameter values $A = 1, B = 0.1, C = 0$, we would have obtained the same sequence $p(n)$ with the exception of the very first sample, $p(O)$, which would have been zero instead of unity.

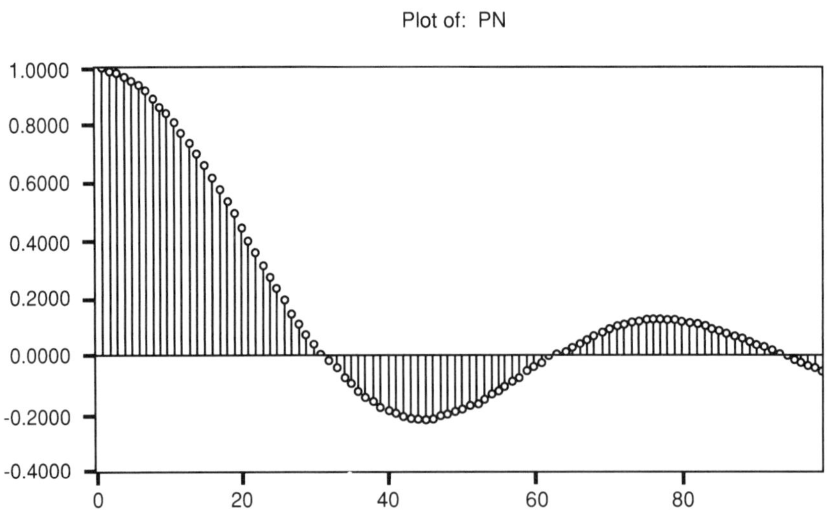

Figure 3-17. The sequence PN of Ex. 3-18.

3-8 PC-DSP DATA FILES

In Sec. 2.2, an example of loading a sequence from a disk data file was presented. In this section, the use of data files is discussed in more detail.

Data files are an important part of PC-DSP. The capability to save and retrieve sequences allows the user to keep the results of his or her work and use them later. Also, data can be transferred between PC-DSP and other programs through the use of data files. A sequence generated within PC-DSP can be saved into a data file to be read by another program. The output of another program can be loaded into PC-DSP for graphing or further processing.

In PC-DSP, each data file contains one sequence. Two data file formats are supported: DSP files and ASCII files.

A DSP file has the default filename extension DSP (e.g., MYDATA1.DSP). These files are unique to PC-DSP and are written to disk as binary files. A DSP file contains all the necessary information about the sequence it stores. In general, it cannot be read by other programs.

An ASCII file, on the other hand, is a straight text file that can be created and/or edited by any ASCII compatible text editor. It can be read by other programs, displayed on the console, or sent to the line printer. It contains only the samples of the sequence it stores. The name and the type of the sequence must be specified by the user after the file is loaded.

Which file format should you use for saving a particular sequence? The answer depends on what you intend to do with that file. A DSP file is significantly smaller than the equivalent ASCII file; therefore, it loads a bit faster. Also, it contains information about the variable name, type, and size. On the other hand, an ASCII file has the advantage of being compatible with other programs.

Example 3-19:

We will now do a *OS-shell* and look at the contents of the ASCII data file MYDATA2.DAT which is on the PC-DSP Utilities and Examples Disk. At the operating system prompt, issue the DOS command

```
TYPE MYDATA2.DAT
```

This command is used for displaying the contents of an ASCII file. The file MYDATA2.DAT has 10 numbers in it. (One number per line).

```
 1.37
-2.29
 4.16
 0.00
 6.03
-1.22
 1.49
 7.52
```

```
 3.26
-0.37
```

Now type EXIT to return to PC-DSP and select the *Load* function from the *System* menu.

```
File type        :  A
Filename [.DAT]  :  MYDATA2
```

Since this is an ASCII file, we will have to specify the variable name.

```
Variable name    :   XA
```

The next prompt is about the type of the variable loaded.

```
Variable type        :  R          R = Real
                                   C = Complex
```

The program responds with

```
Variable = XA        , Type = R , Size = 10
```

after loading the 10-point sequence into the variable XA.

Example 3-20:

In this example, we will load the same file again, but this time we will load it as a complex sequence. Select *Load*, and answer prompts as indicated:

```
File type          :  A
Filename [.DAT]    :  MYDATA2
Variable name      :  XB
Variable type      :  C
```

The program now responds with

```
Variable = XB        , Type = C , Size = 5
```

If the variable type is specified as complex, the numbers in the file are loaded in sets of 2. The first number in each set is taken as the real part of the corresponding sample, and the second number becomes the imaginary part. You can use the *Print* function of the *System* menu to verify that the loaded sequence XB has the following five complex samples:

$$x_b(0) = 1.37 - j2.29$$

$$x_b(1) = 4.16 + j0.00$$

$$x_b(2) = 6.03 - j1.22$$

$$x_b(3) = 1.49 + j7.52$$

$$x_b(4) = 3.26 - j0.37$$

3-9 EXERCISES

1. Generate and graph each one of the following sequences for $n = 0, \ldots, 99$.
   a. $\delta(n-3) - 2\delta(n-5)$
   b. $u(n-5)$
   c. $\sin(0.2n) + w(n)$, where $w(n)$ is Gaussian white noise with zero mean and unit variance
   d. $\exp[(-0.2 + j0.3)n]$
   e. $n[u(n) - u(n-10)] + \exp(-0.3n)[u(n-10) - u(n-20)]$

2. Generate and graph the sequence

$$x(n) = (0.9)^n \sin(0.2n) \qquad \text{for } n = 0, \ldots, 99$$

Hint: Although this sequence can be generated by generating two sequences $(0.9)^n$ and $\sin(0.2^n)$ separately and multiplying them, you can save some steps if you realize that

$$x(n) = \text{Imag}\{(0.9e^{j0.2})n\}$$

3. Generate six periods of a periodic pulse train with each period consisting of 15 ones followed by 10 zeros.

4. Using the same message and carrier signals used in Ex. 3-17, generate and graph the double sideband AM signal with *suppressed carrier,* i.e., generate

$$x_{SC}(n) = x_m(n)x_c(n), \qquad \text{for } n = 0, ..., 256$$

4

Basic Signal Processing Functions

In this chapter, we take a look at the basic signal processing functions that are found in the *Data* component of PC-DSP. These functions are in the *Process* and *Transform* submenus of the *Data* component. They implement forward and inverse discrete Fourier transforms, convolution, correlation, inverse z-transform, decimation, zero-padding, and zero-insertion operations. We give a brief mathematical description of each function and then illustrate its use with example problems. After you go through each section, feel free to explore further the functions covered.

4-1 DISCRETE FOURIER TRANSFORM

The Discrete Fourier Transform (DFT) of an N-point sequence $x(n)$ is defined by the summation

$$X(k) = \sum_{n=0}^{N-1} x(n) \exp\left(-j\frac{2\pi nk}{N}\right) \qquad \text{for } k = 0, \ldots, N-1 \tag{4-1}$$

If the transform $X(k)$ is given instead, then the inverse discrete Fourier transform can be computed as

$$x(n) = \frac{1}{N} \sum_{k=0}^{N-1} X(k) \exp\left(j\frac{2\pi nk}{N}\right) \qquad \text{for } n = 0, \ldots, N-1 \tag{4-2}$$

In PC-DSP, Eqs. (4-1) and (4-2) are implemented by two built-in functions, *DFT* and *iNv-DFT*, respectively. Both functions are in the *Transform* submenu of the *Data* component. Both accept one sequence as input and produce the transform or the inverse transform sequence in return.

Example 4-1:

A sequence $x(n)$ is defined as

$$x(n) = \begin{cases} 1, & n = 0, 1, \ldots, 5 \\ 0, & n = 6, 7, \ldots, 11 \end{cases}$$

Compute the discrete Fourier transform $X(k)$. Plot its magnitude and phase.

First, we need to generate the 12-point sequence $x(n)$. The easiest way to do this is to use the *Read* function of the *System* menu. Select *Read,* and answer the program prompts as follows:

```
Variable name :  XN
Sample values :  1;1;1;1;1;1;0;0;0;0;0;0
```

You may want to glance through Sec. 3-1 if you are not very clear on doing this. If all is well, you should get the message

```
Variable = XN        , Type = R , Size = 12
```

Now select *Data,* then *Transform,* and then *DFT.* The format statement for the *DFT* function appears at the top of the data entry window:

```
FORMAT:  Var1 = DFT { Var2 }
```

In our example, Var2 is the sequence XN that we have just generated, and Var1 is the sequence XK that will be generated by the program as the result of the DFT operation. Fill in the variable names as follows:

```
Var1  :  XK
Var2  :  XN
```

The program responds with

```
Variable = XK        , Type = C , Size =  12
```

signaling the addition of a new sequence to the contents of data memory.

Note that XK is a complex sequence, as expected from Eq. (4-1). If you pressed the [F2] function key now, hoping to get a quick-plot of XK, you would get a plot of XN instead. That is because PC-DSP cannot plot a complex sequence, and the most recently accessed plottable sequence is still XN. To get a graphical view of the transform just computed, we will compute the magnitude and the phase of XK, and plot them instead.

First, select the *Magnitude* function of the *Nonlinear* submenu.

```
FORMAT:   Var1 = | Var2 |
Var1  :    XKM
Var2  :    XK
```

The computed sequence XKM is real, and the quick-plot given in Fig. 4-1 can be obtained with the [F2] key. Now select the *Phase* function of the *Nonlinear* submenu.

```
FORMAT:   Var1 = PHASE { Var2 }
Var1  :    XKP
Var2  :    XK
```

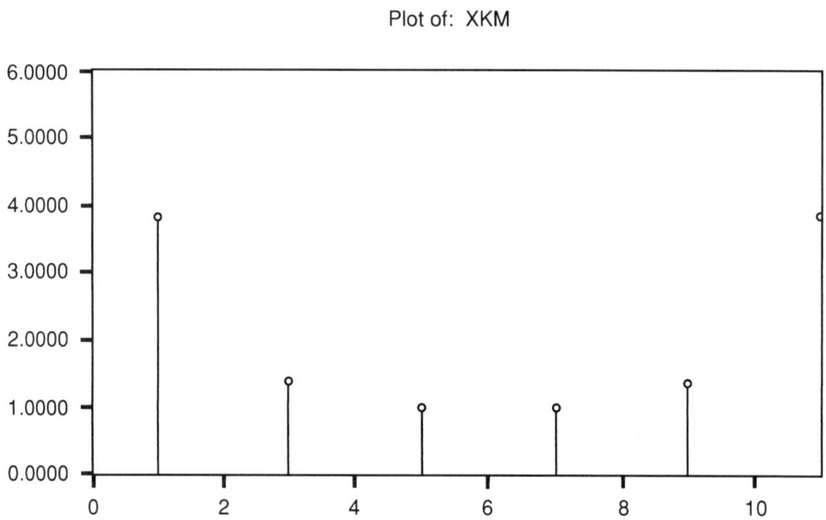

Figure 4-1. Magnitude plot for Ex. 4-1.

The phase sequence XKP contains four-quadrant radian phase angles of the samples of XK. If desired, the phase can be converted to degrees by scaling the sequence XKP with a constant scale factor $180/\pi$. Select *Arithmetic,* then *Multiply,* then *Con*seq.*

```
FORMAT:   Var1 = Constant * Var2
Var1  :    XKP
Var2  :    XKP
Const :    180/π
```

Note that variable names Var1 and Var2 are both specified as XKP. The input sequence in XKP is multiplied by the constant scale factor $180/\pi$, and the result is written into XKP, overwriting the input sequence. Thus, XKP now contains phase angles of the transform samples in degrees. The quick-plot is shown in Fig. 4-2.

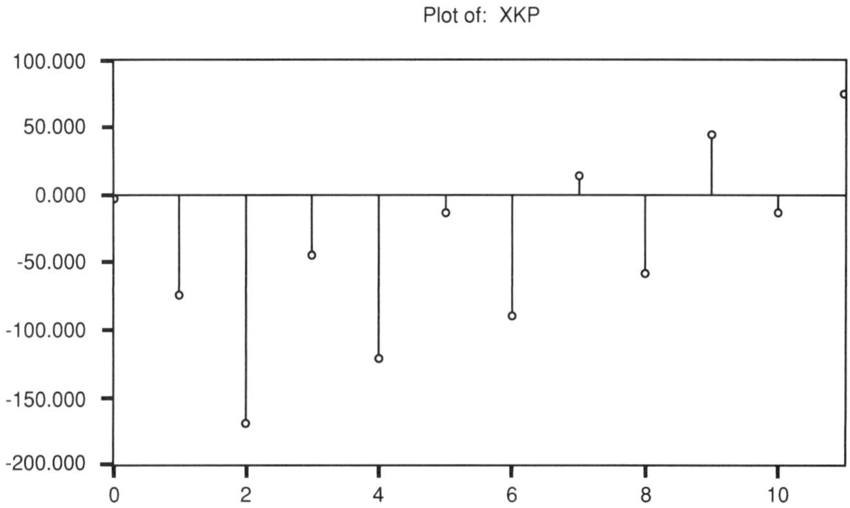

Figure 4-2. Phase plot (in degrees) for Ex. 4-1.

4-2 FAST FOURIER TRANSFORM

The *Transform* submenu of the *Data* component also contains functions for forward and inverse fast Fourier transform. These functions are *FFT* and *Inv-FFT*, respectively. They implement Eqs. (4-1) and (4-2), the same equations that are implemented by functions *DFT* and *iNv-DFT*. The only difference is that *DFT* and *iNv-DFT* implement the summations in (4-1) and (4-2) directly, whereas *FFT* and *Inv-FFT* functions use radix-2 decimation-in-time algorithms to perform the computations much more quickly. The fast Fourier transform of an N-point sequence requires about $N \log_2 N$ complex additions and multiplications compared with N^2 for a direct implementation of the summation.

In general, both the input and the output sequences involved may be complex and up to 4096 samples long. The only other requirement is that the transform size N be an integer power of 2. This requirement is due to the radix-2 decimation-in-time algorithm used by the program and does not pose any problems. If the length of the sequence to be transformed is not an integer power of 2, then the program internally generates an "extended sequence" by padding the original sequence with zeros to extend its length to the nearest integer power of 2 and then computes the transform (or the inverse transform as the case may be) of this extended sequence. For example, if you attempted to compute the fast Fourier transform of the five-sample sequence

$$x(n) = \{3, 2, -1, 1, 6\}$$

the program would actually compute the eight-point transform of the extended sequence

$$\tilde{x}(n) = \{3, 2, -1, 1, 6, 0, 0, 0\}$$

which has three zeros appended to it to make it eight samples long. The length of the original sequence $x(n)$ does not change. The extended sequence $\tilde{x}(n)$ is discarded after the transform is computed.

Example 4-2:

The discrete Fourier transform computed in Ex. 4-1 was 12 samples long. Sometimes it is desirable to see the frequency spectrum of a signal in more detail. One way to obtain more detail in the transform is to pad the sequence with a set of zero-amplitude samples. In this example, we will pad the sequence XN with 52 zeros to extend its length to 64. This will allow us to get a 64-point transform when the *FFT* function is used.

The *Process* submenu of the *Data* component has a function called *zero-Pad* that can be used for padding a sequence with a desired number of zeros. The format is

```
FORMAT :  Var1 = ZERO-PAD { Var2 , #Zeros }
```

The output sequence Var1 is generated by appending Var2 with as many zeros as specified by the integer variable #Zeros. We will call the new sequence XNL.

```
Var1    :   XNL
Var2    :   XN
#Zeros  :   52
```

XNL is 64 samples long. The magnitude XKML and the phase XKPL of the discrete Fourier transform of this sequence can be obtained by following the same procedure used in Ex. 4-1. Quick-plots of XKML and XKPL are given in Figs. 4-3 and 4-4.

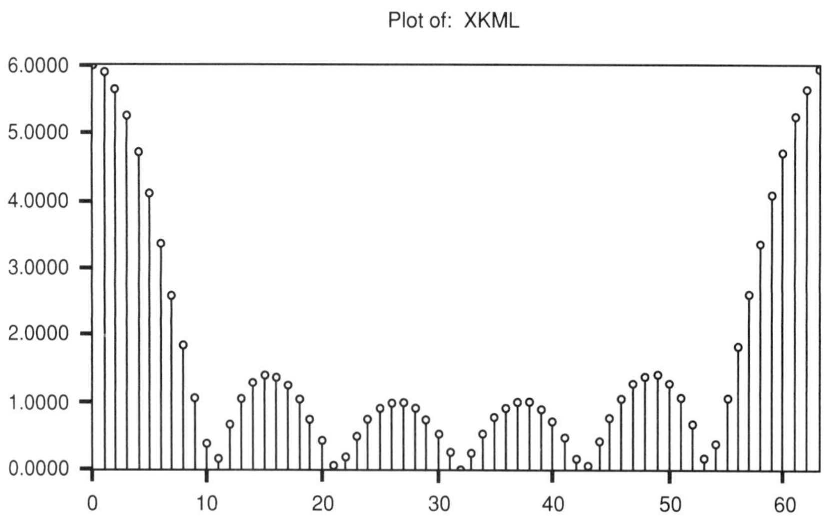

Figure 4-3 Magnitude plot for Ex. 4-2.

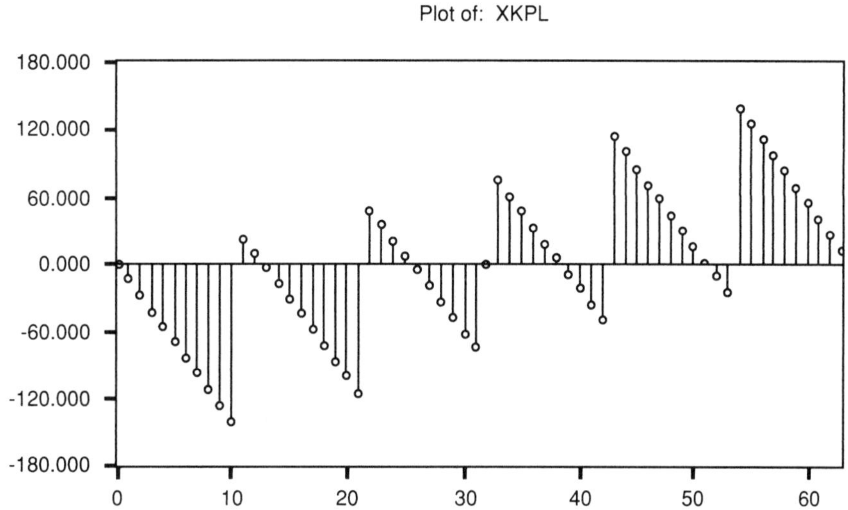

Figure 4-4 Phase plot (in degrees) for Ex. 4-2.

Example 4-3:

Compute the frequency spectrum of the amplitude modulated signal XAMN of Ex. 3-17.

If the sequence XAMN is no longer in memory, it can be loaded from the data file EX3-17.DSP. We will first compute the FFT and then take its magnitude. Select *Transform*, then *FFT*.

```
FORMAT :  Var1 = FFT { Var2 }
Var1   :    XAMK
Var2   :    XAMN
```

Now, select *Nonlinear*, then *Magnitude*.

```
FORMAT :  Var1 = | Var2 |
Var1   :    XAMK
Var2   :    XAMK
```

A quick-plot of XAMK is given in Fig. 4-5. The 256 samples shown represent samples of the magnitude spectrum of XAMN taken every $2\pi/256$ radians. Note that the sidebands centered around the carrier frequency are not clearly visible. This is due to the fact that the message frequency is very low, and the sidebands are very close to the carrier frequency. We are just not getting enough samples of the magnitude spectrum to show the detail around the carrier. By padding the sequence XAMN with 768 zeros to extend its length to 1024 and then computing a 1024-point FFT, we can obtain four times as many samples of the magnitude spectrum, with transform samples spaced $2\pi/1024$ radians apart. Now select the *zero-Pad* function from the *Process* submenu.

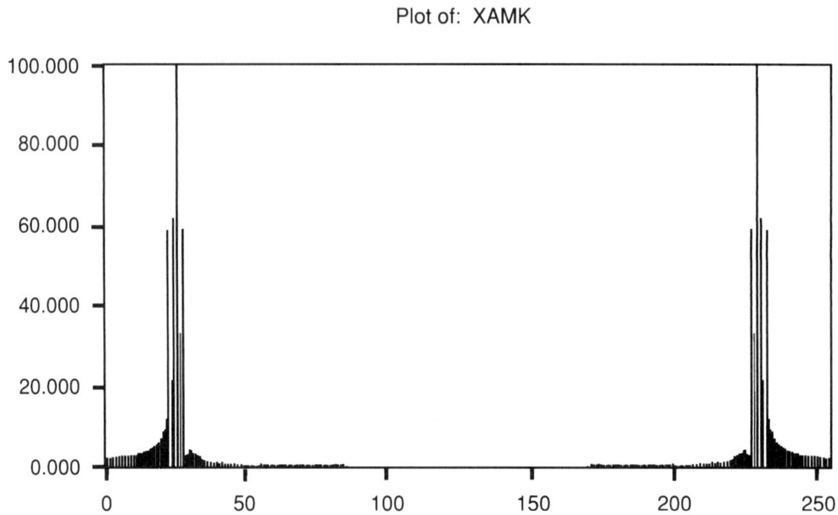

Figure 4-5 256-point magnitude spectrum obtained in Ex. 4-3.

```
FORMAT:  Var1 = ZERO-PAD { Var2, #Zeros }
Var1   :  XAMNL
Var2   :  XAMN
#Zeros:   768
```

Next perform an *FFT* on this extended sequence.

```
FORMAT:  Var1 = FFT { Var2 }
Var1   :  XAMKL
Var2   :  XAMNL
```

Select the *Magnitude* function from the *Nonlinear* submenu.

```
FORMAT:  Var1 = | Var2 |
Var1   :  XAMKL
Var2   :  XAMKL
```

The new magnitude spectrum, XAMKL, is 1024 samples long. Since the second half of it is a mirror image of the first half, we will use the *Copy* function of the *Edit* submenu to retain the first 512 samples of XAMKL.

```
FORMAT:  Var1 = COPY { Var2, From, To }
Var1   :  XAMKL
Var2   :  XAMKL
From   :  0
To     :  511
```

A quick-plot of XAMKL is shown in Fig. 4-6. In Chapter 5, we will customize this graph by scaling the horizontal axis to show the actual radian frequencies instead of sample numbers. That will allow us to see more clearly the relation between peaks of the spectrum and frequencies used in Ex. 3-17 for carrier and message signals.

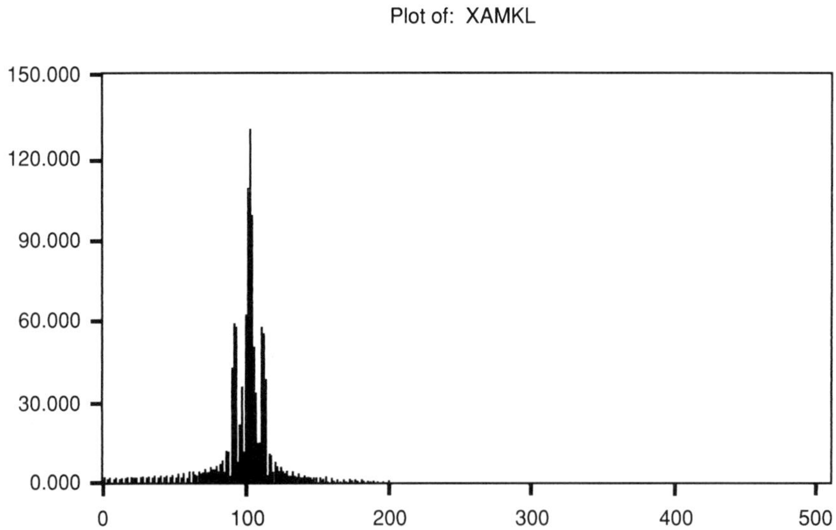

Figure 4-6 Magnitude spectrum obtained in Ex. 4-3.

4-3 CONVOLUTION

Linear convolution is an operation that is used very often in linear system problems involving the computation of the output sequence of a linear system driven by a specified input sequence. Consider a linear time-invariant system characterized by its impulse response $h(n)$. If the system is driven by an input sequence $x(n)$, the output sequence can be found using the well-known convolution sum

$$y(n) = \sum_k h(k)x(n-k) \tag{4-3}$$

where the range of the summation index k should include all overlapping samples of $h(k)$ and $x(n-k)$. The *Convolve* function of the *Process* submenu implements the convolution sum of Eq. (4-3) in a fast way. The format is

```
FORMAT:  Var1 = CONVOLVE { Var2, Var3 }
```

The two input sequences, Var2 and Var3, are convolved, and the result is stored in Var1. If the lengths of input sequences are N_2 and N_3, respectively, then the output sequence Var1 is $N_2 + N_3 - 1$ samples long. This length must not exceed 4096 samples.

Example 4-4:

The impulse response of a linear time-invariant system is

$$h(n) = \begin{cases} e^{-0.15n} & n = 0, \ldots, 31 \\ 0 & \text{otherwise} \end{cases}$$

This system is driven by the input sequence

$$\text{input}(n) = u(n) - u(n - 20)$$

where $u(n)$ denotes the unit-step. Compute the output sequence using convolution.

The impulse response, $h(n)$, of the system can be generated using the *Exp* function of the *Generate* submenu.

```
Var1 = A EXP { B n + C }  ,   n = 0, ...#Sampl-1
Var1   :  HN
  A    :  1
  B    :  -0.15
  C    :  0
#Sampl:   32
```

Thus, we have obtained the impulse response of the system. To obtain the input sequence, we will use the *Ramp* function of the *Generate* submenu. Recall that a step sequence is basically a ramp sequence with zero slope.

```
Var1 = Init + n.Incr   ,   n = 0, ..., #Sampl-1
Var1   :    INPUT
Init   :    1
Incr   :    0
#Sampl:     20
```

We are now ready to convolve the sequences HN and INPUT. Select *Process*, and then select *Convolve*.

```
FORMAT:  Var1 = CONVOLVE { Var2, Var3 }
Var1   :   OUTPUT
Var2   :   INPUT
Var3   :   HN
```

The output sequence OUTN is 51 samples long, as shown in Fig. 4-7.

The *Convolve* function implements the linear convolution sum. Circular convolution is not directly implemented in PC-DSP, but it can be computed using the functions *DFT* and *iNv-DFT*. It is also possible to write a macro file to do circular convolution. We will look at macro files in detail in Chapter 9. In that chapter, we will also consider adding a circular convolution function to PC-DSP by means of a macro file.

4-4 AUTO- AND CROSS-CORRELATION

The *Auto-corr* function of the *Process* submenu is used in computing an estimate of the autocorrelation of a sequence. If $x(n)$ is an infinite-length power sequence, its autocorrelation is defined as

$$\phi_{xx}(m) = < x^*(n)x(n+m) > \tag{4-4}$$

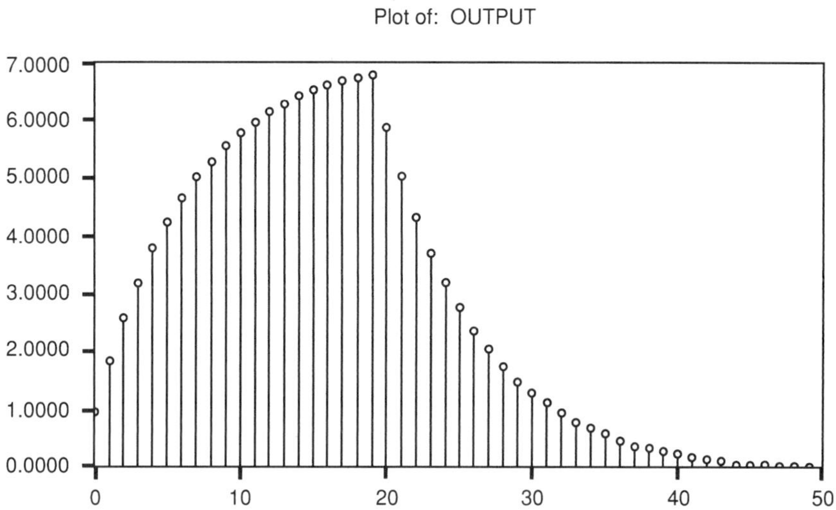

Figure 4-7. The output sequence for Ex. 4-4.

where the angle brackets $< \ldots >$ indicate averaging over the time variable n and the asterisk denotes complex conjugate. The result, $\phi_{xx}(m)$, is also an infinite-length sequence. This definition of the autocorrelation sequence is based on the assumption that the time domain sequence $x(n)$ satisfies a condition called *ergodicity*. Theoretical details of this subject are beyond the scope of this workbook. They can be found in most digital signal processing textbooks.

If we have a finite segment of data from the infinite-length sequence $x(n)$, we can use that finite segment to *estimate* the autocorrelation. Let $x_F(n)$ be an N-sample segment of $x(n)$. We will use the estimation formula known as the *biased estimator*.

$$r_{xx}(m) = \frac{1}{N} \sum_{n} x_F^*(n) x_F(n+m) \qquad \text{for } m = 0, \ldots, N-1 \tag{4-5}$$

where the summation is carried out for a range of the index n that includes all overlapping samples of $x_F(n)$ and $x_F(n+m)$.

Example 4-5:

White noise is known to be the most unpredictable random signal. An indication that a signal is unpredictable is an autocorrelation function in the shape of an impulse. A normally distributed pseudo-random sequence is often used as an approximation to white noise in computer simulation problems. In this example, we will obtain the autocorrelation estimate of the sequence NOISE2 generated in Ex. 3-12 and see if it really approximates an impulse sequence. If you no longer have the sequence NOISE2 in memory, it can be loaded from the data file EX3-12.DSP. Select *Process*, then *Auto-corr*.

```
FORMAT:  Var1 = AUTO-CORR { Var2 }
Var1   :     RM
Var2   :     NOISE2
```

The autocorrelation sequence RM is shown in Fig. 4-8. Even though it is not an ideal impulse, it is close. You may want to try increasing the number of samples in the random sequence NOISE2 and repeating this example. A longer sequence should yield an autocorrelation estimate that better approximates an impulse.

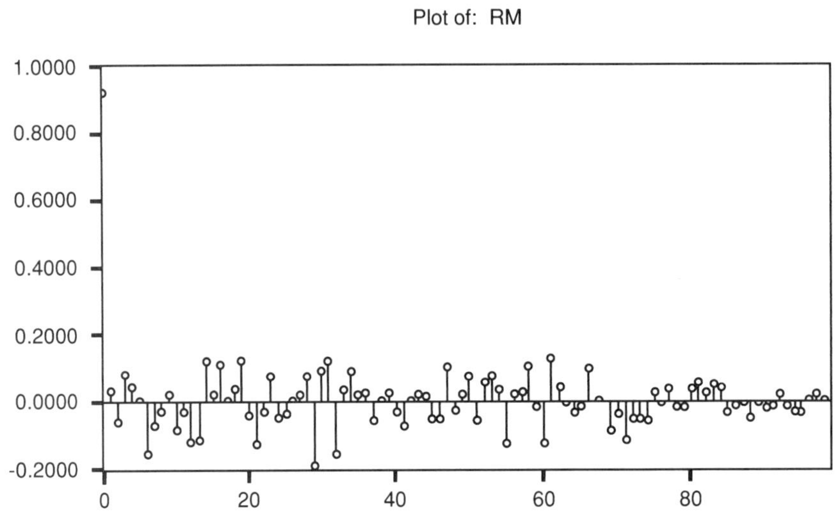

Figure 4-8 Autocorrelation estimate of NOISE2.

Example 4-6:

Another application of the autocorrelation function is in detecting periodicities in a noisy signal. Consider an infinite-length signal, $y(n)$, that consists of a single tone buried in white noise, i.e.,

$$y(n) = \sin(0.5n) + v(n)$$

where $v(n)$ represents a zero-mean unit-variance white noise process that is independent of the signal component $\sin(0.5n)$. The autocorrelation of $y(n)$ has two components. The component due to $v(n)$ is an impulse, as we have seen in Ex. 4-5, and it can be shown that the component due to $\sin(0.5n)$ is $0.5 \cos(0.5m)$. The autocorrelation of $y(n)$ can be written as

$$\phi_{yy}(m) = 0.5\cos(0.5m) + \delta(m)$$

If we gathered a finite segment of data from the infinite-length signal $y(n)$ just given and computed the autocorrelation estimate $r_{yy}(n)$, we would expect to see an approximation to the ideal autocorrelation sequence $\phi_{yy}(m)$. Such a finite segment of data has been stored in the data file EX4-6.DSP. Load the file to obtain the 100-point sequence YFN (see Fig. 4-9). At first glance, it is quite difficult to recognize the sinusoidal component. Select *Process* and then *Auto-corr*.

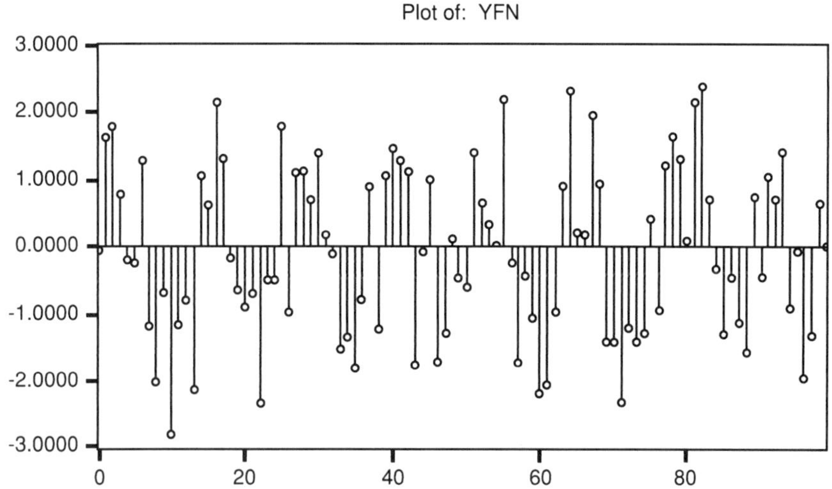

Figure 4-9 The sequence YFN of Ex. 4-6.

```
FORMAT:  Var1 = AUTO-CORR { Var2 }
Var1  :     RYM
Var2  :     YFN
```

Autocorrelation estimate RYM is shown in Fig. 4-10. The periodic component is clearly evident.

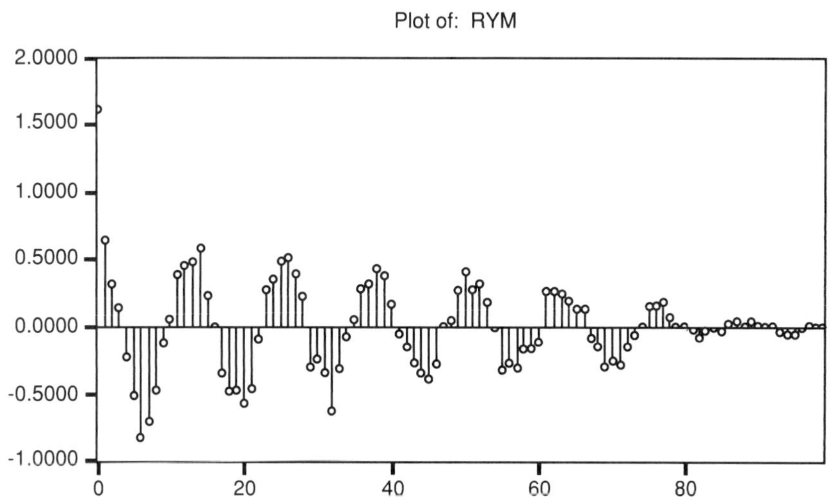

Figure 4-10 Autocorrelation estimate for YFN.

The cross-correlation of two infinite-length power sequences $x(n)$ and $y(n)$ is defined as

$$\phi_{xy}(m) = < x^*(n)y(n+m) > \tag{4-6}$$

where the angle brackets indicate averaging over the time variable. In this definition, it is again inherently assumed that the sequences involved are ergodic.

If finite data segments are available from the infinite-length sequences $x(n)$ and $y(n)$, they can be used to estimate the cross-correlation sequence. Let $x_F(n)$ and $y_F(n)$ be N-sample segments of $x(n)$ and $y(n)$, respectively. The *biased estimator* formula

$$r_{xy}(m) = \frac{1}{N}\sum_n x_F^*(n)y_F(n+m) \tag{4-7}$$

can be used to find an estimate $r_{xy}(m)$ of $\phi_{xy}(m)$. The summation in Eq. (4-7) is carried out over the entire range of overlapping samples of $x_F(n)$ and $y_F(n+m)$. The *cRoss-corr* function of the *Process* submenu implements such an estimator. The format of this function is

```
FORMAT:  Var1 = CROSS-CORR { Var2, Var3 }
```

Example 4-7:

An interesting application of cross-correlation is in estimating the impulse response of an unknown system, driven by white noise. Let us consider a linear system with unknown impulse response $h(n)$. If the input sequence $x(n)$ is white noise with zero mean and unit variance, then it can be shown that the cross-correlation of the input and the output sequences is equal to the unknown impulse response, i.e.,

$$\phi_{xy}(m) = h(m)$$

The impulse response of such a linear system is stored in the data file EX4-7.DSP. Load the file to obtain the sequence HN. You can look at the impulse response just loaded by pressing the [F2] function key (Fig. 4-11).

Let's pretend now that we don't know the impulse response of the system. We will use the normal random number generator to generate 300 samples of an approximate white noise sequence. (We will use a relatively large number of samples so that a good estimate of the cross-correlation sequence can be obtained.) Select *Generate*; then select *Normal*.

```
Var1 = NORMAL { Mean, Variance, #Sampl }
Var1      :     XN
Mean      :     0
Variance:       1
#Sampl    :     300
```

The sequence XN will be used as input to the unknown system. The easiest way to compute the output sequence YN is to convolve XN and HN using the *Convolve* function.

```
FORMAT:  Var1 = CONVOLVE { Var2, Var3 }
Var1  :     YN
```

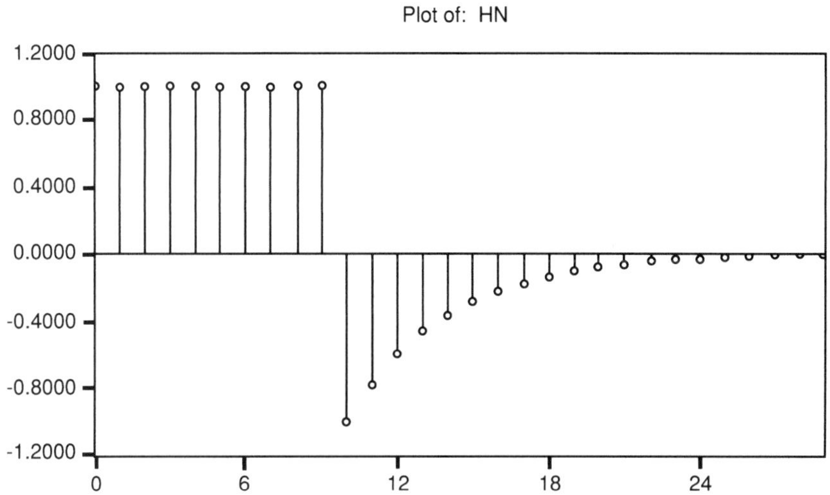

Figure 4-11 Impulse response HN for the system in Ex. 4-7.

```
Var2  :   XN
Var3  :   HN
```

Now select *cRoss-corr* to compute the cross-correlation estimate for XN and YN.

```
FORMAT:  Var1 = CROSS-CORR { Var2, Var3 }
Var1  :   RXYM
Var2  :   XN
Var3  :   YN
```

A quick-plot of the cross-correlation estimate RXYM is shown in Fig. 4-12. The cross-correlation estimate is 329 samples long, but we are interested in only the first few samples. Extract samples 0 through 28 using the *Copy* function of the *Edit* submenu.

```
FORMAT:   Var1 = COPY { Var2, From, To }
Var1  :   RXYM
Var2  :   RXYM
From  :   0
To    :   28
```

The new sequence RXYM is shown in Fig. 4-13. Compare RXYM with the true impulse response of the system, shown in Fig. 4-11.

4-5 INVERSE Z-TRANSFORM

The *inv-Z* function of the *Transform* submenu can be used for computing the inverse z-transform of a rational function through long division. Its format is

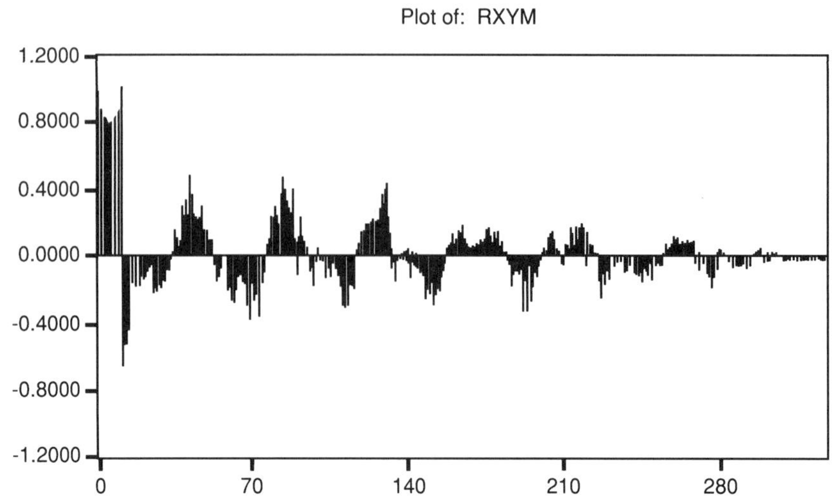

Figure 4-12 329-sample cross-correlation estimate of Ex. 4-7.

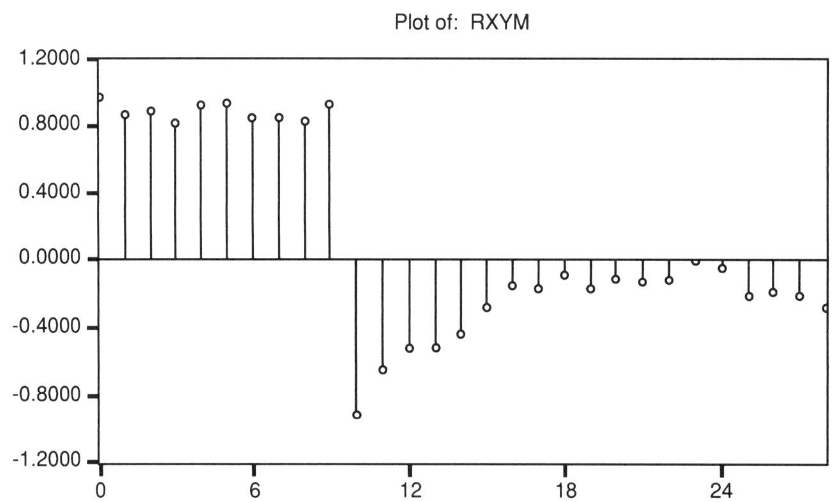

Figure 4-13 The first 29 samples of the cross-correlation estimate of Ex. 4-7.

```
FORMAT:  Var1 = INV-Z { Num, Den, #Sampl }
```

where the arguments are named sequences holding the numerator and the denominator coefficients, respectively. These two sequences should be created prior to using the *inv-Z* function. In most cases, the easiest way to create the numerator and denominator sequences is to use the keyboard entry method.

Example 4-8:

Compute 100 samples of the impulse response of a system described by the z-domain transfer function

$$H(z) = \frac{z^2 + 1}{z^3 - 1.9z^2 + 1.55z - 0.425}$$

We will first need to generate two variables to hold numerator and denominator polynomial coefficients. Select the *Read* function of the *System* menu, and enter the following two sequences.

```
HNUM = {1  0  1}
HDEN = {1  -1.9  1.55  -0.425}
```

Note that the numerator and denominator coefficients are entered from the highest positive power of z to the lowest. Now select *Data*, then *Transform*, then *inv-Z*.

```
FORMAT:   Var1 = INV-Z { Num, Den, #Sampl }
Var1   :   HN
Num    :   HNUM
Den    :   HDEN
#Sampl:    100
```

The first 100 points of the impulse response of the given system are now stored in HN. The quick-plot is shown in Fig. 4-14.

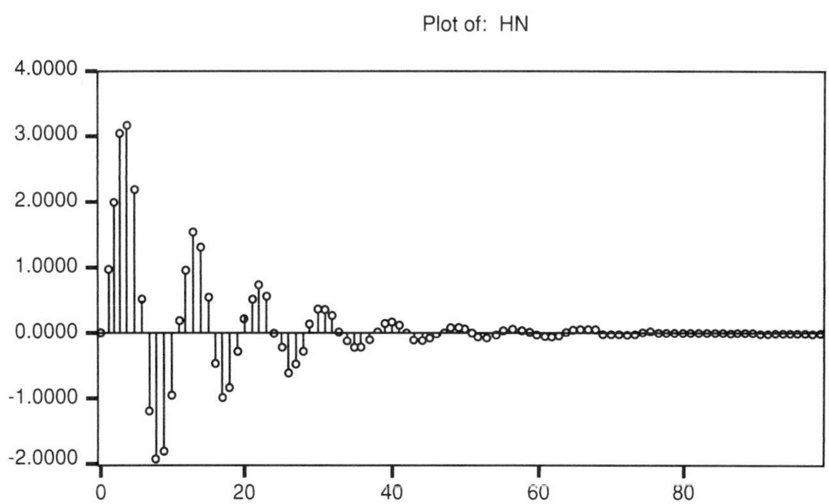

Figure 4-14 The impulse response of the system in Ex. 4-8.

Example 4-9:

Compute the first 100 samples of the unit-step response of the system given in Ex. 4-8.
The input x(n) is a unit-step with the z-transform

$$X(z) = \frac{z}{z-1}$$

and the output z-transform $Y(z)$ is the product of the input z-transform and the system function, i.e.,

$$Y(z) = H(z)X(z)$$

We can obtain numerator and denominator polynomials of $Y(z)$ and then use *inv-Z* to find the output sequence $y(n)$ as the inverse z-transform of $Y(z)$. Let's use the *Read* function to type in the following two sequences:

```
XNUM = {1   0}
XDEN = {1  -1}
```

We now have the numerator and the denominator coefficients of $X(z)$ in the sequences XNUM and XDEN, respectively. Numerator and denominator polynomials of $Y(z)$ can now be computed as products of corresponding polynomials in $H(z)$ and $X(z)$. The polynomial multiplications required can be carried out by hand, but we will be a little more creative and use the *Convolve* function to multiply two polynomials.

```
FORMAT:    Var1 = CONVOLVE { Var2, Var3 }
Var1  :     YDEN
Var2  :     HDEN
Var3  :     XDEN
```

You may want to do this polynomial multiplication with pencil and paper to convince yourself that this method really works. The numerator polynomial YNUM can be obtained in the same way by convolving HNUM and XNUM. Samples of the output sequence YN can then be computed using the *inv-Z* function (see Fig. 4-15).

4-6 DECIMATION AND INTERPOLATION

Sampling rate conversion, i.e., changing the sampling rate of a sequence by a desired factor, is an important operation in digital signal processing. In the *Process* submenu of the *Data* component, two functions are provided for this purpose, *Decimate* and *Zero-ins*.

A decimator reduces the sampling rate of a sequence by an integer factor. This is sometimes called the *down-sampling operation*. If the input sequence $x_{in}(n)$ is to be decimated by a factor of M, the output sequence of the decimator is computed as

$$x_{out}(n) = x_{in}(nM)$$

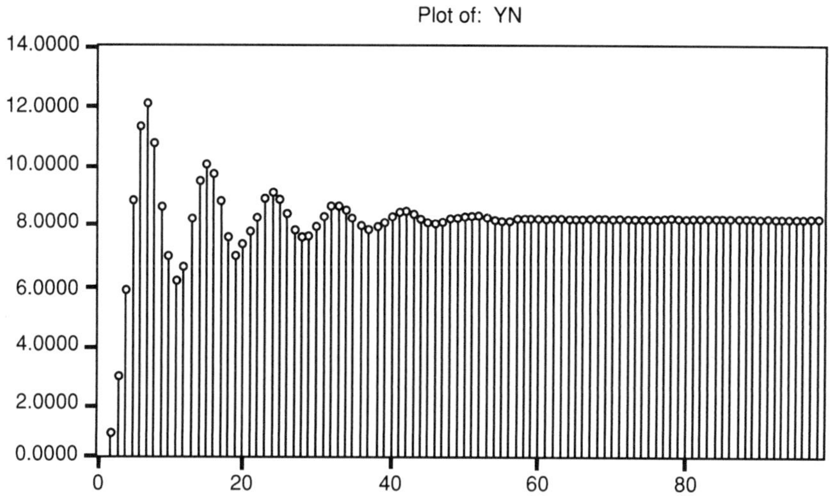

Figure 4-15. Unit-step response of the system of Ex. 4-9.

Thus, the output sequence contains every Mth sample of the input sequence. We illustrate the *Decimate* function with a simple example.

Example 4-10:

Generate 40 samples of the sinusoid

$$w(n) = \sin(0.25n), \qquad n = 0, \ldots, 39$$

Decimate $w(n)$ with a decimation factor of 2 to obtain a 20-sample sequence $w_1(n)$.

The sinusoid $w(n)$ can be generated using the *Sine* function of the *Generate* submenu with parameter values $A = 1, B = 0.25, C = 0$, and #Sampl = 40, as explained in Chapter 3. After generating $w(n)$, select the *Decimate* function. Its format is

```
FORMAT:  Var1 = DECIMATE { Var2, Factor }
```

where Var1 is obtained by decimating Var2 with the specified decimation factor. Fill in the variable names and the decimation factor as follows:

```
Var1   :    W1N
Var2   :    WN
Factor:     2
```

Quick-plots for WN and W1N are given in Figs. 4-16 and 4-17. The output sequence W1N contains every other sample in WN. In a practical decimator, the input sequence would first be processed through a band-limiting filter to avoid the aliasing phenomenon. We will leave this to Chapter 6, where design and analysis of FIR filters are discussed.

Sometimes it is desired to increase the sampling rate of a sequence, an operation called *interpolation*, or *up-sampling*. The *Zero-ins* function inserts zeros between samples

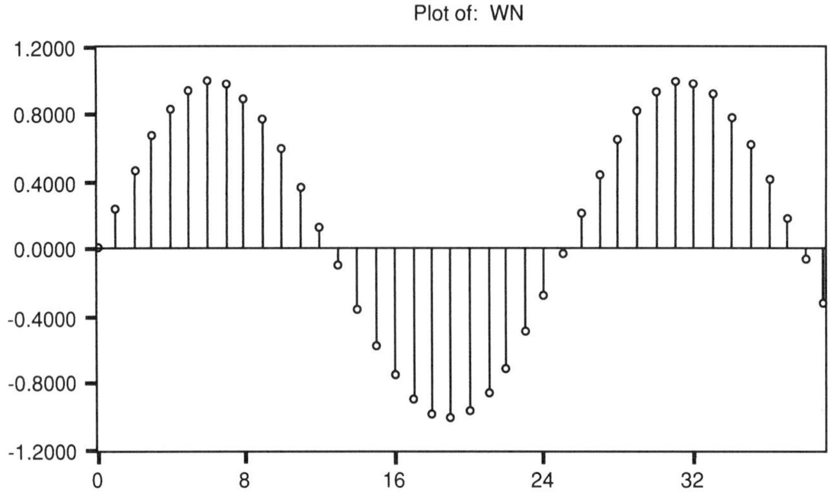

Figure 4-16. The sequence WN of Ex. 4-10.

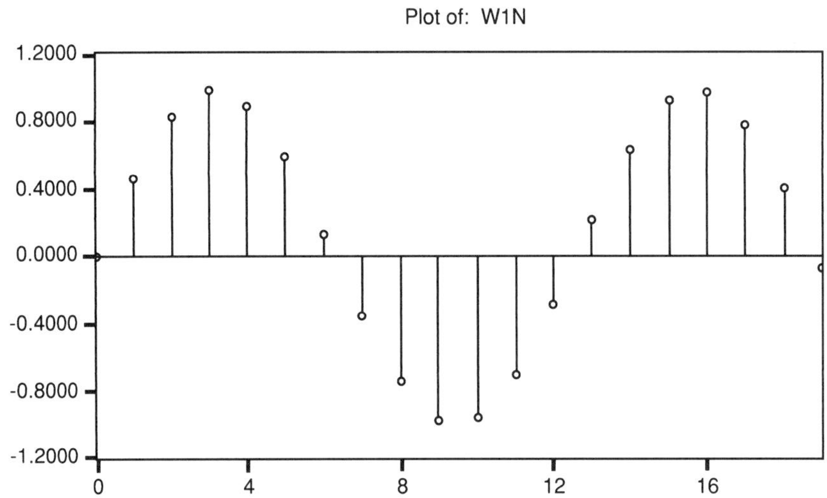

Figure 4-17. The sequence W1N of Ex. 4-10.

of a sequence. This is typically the first step in the up-sampling operation. The use of the *Zero-ins* function with a factor of M results in the output sequence

$$x_{out}(n) = \begin{cases} x_{in}(n/M), & n = kM \\ 0, & \text{otherwise} \end{cases}$$

where $M - 1$ zeros are inserted after each sample of $x_{in}(n)$. After the zero insertion, an interpolator filter can be used to change the zeros to interpolated values.

Example 4-11:

Double the sampling rate of the sequence W1N obtained in Ex. 4-10, by first using zero insertion with a factor of 2 and then processing the sequence through a zero-order-hold interpolator.

Select the *Zero-ins* function, and make the following entries:

```
FORMAT:  Var1 = ZERO-INS { Var2, Factor }
Var1   :    W2N
Var2   :    W1N
Factor:     2
```

The output sequence W2N contains 40 samples where every other sample is a zero (see Fig. 4-18).

A zero-order-hold interpolator can be implemented as an FIR filter that is described by the difference equation

$$y(n) = x(n) + x(n-1)$$

It can be shown that the impulse response of this FIR filter is

$$h(n) = \begin{cases} 1, & n = 0, 1 \\ 0, & \text{otherwise} \end{cases}$$

Generate the impulse response HN using the *Read* function. Convolve W2N and HN to obtain the sequence W3N (see Fig. 4-19).

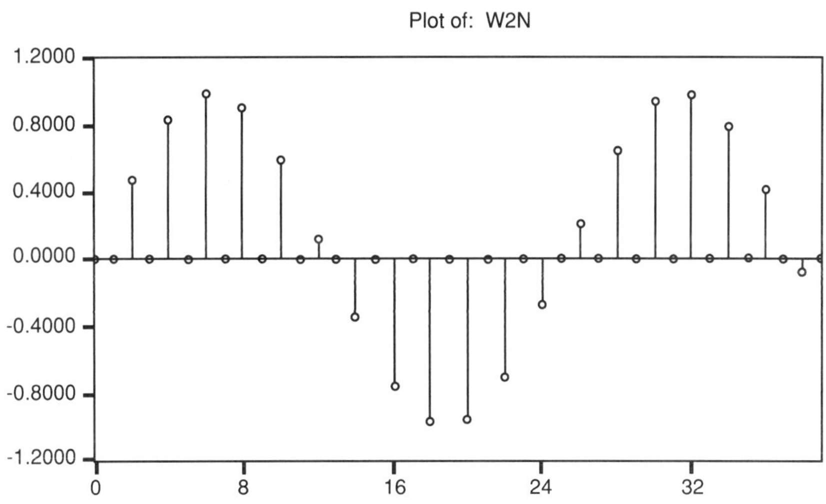

Figure 4-18. The sequence W2N of Ex. 4-11.

Example 4-12:

Repeat Ex. 4-11 using a first-order-hold (linear) interpolator described by the difference equation

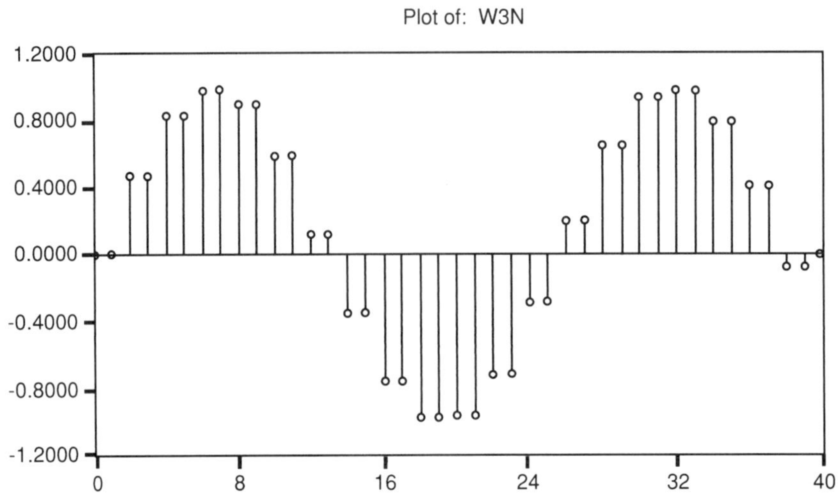

Figure 4-19. The sequence W3N of Ex. 4-11.

$$y(n) = 0.5x(n-1) + x(n) + 0.5x(n+1)$$

The impulse response of this first-order-hold interpolator is

$$h(n) = \begin{cases} 0.5, & n = \mp 1 \\ 1, & n = 0 \\ 0, & \text{otherwise} \end{cases}$$

Apparently, $h(n)$ is noncausal. To simulate a system that has M noncausal samples, we will add M units of delay and then M units of advance (see Fig. 4-20).

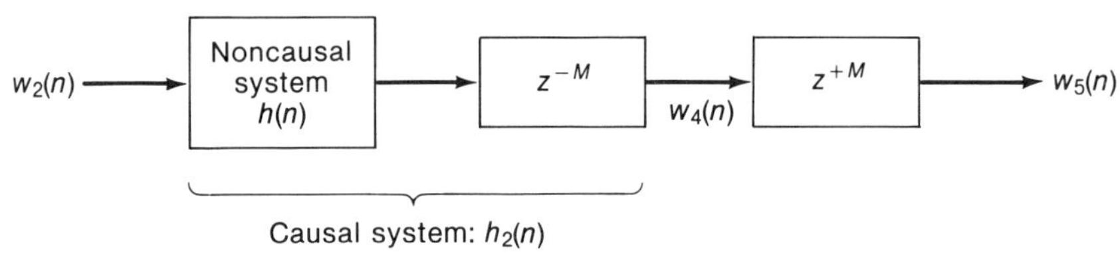

Figure 4-20. Simulating a noncausal system.

Generate H2N by selecting the *Read* function and entering sample values 0.5; 1.0; 0.5 for n = 0, 1, 2. W4N can now be obtained by convolving H2N with W2N.

```
FORMAT:  Var1 = CONVOLVE { Var2, Var3 }
Var1  :   W4N
Var2  :   W2N
Var3  :   H2N
```

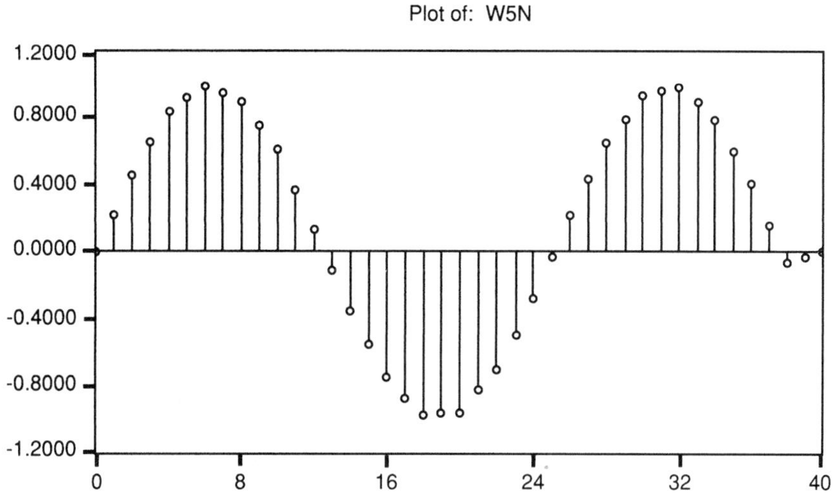

Figure 4-21. The sequence W5N of Ex. 4-12.

To obtain W5N, we need to shift W4N to the left by one sample. (The impulse response was shifted to the right by one sample to make the system causal.) A left-shift can be achieved by using a negative number with the *Shift* function.

```
FORMAT:  Var1 = SHIFT { Var2, #Sampl }
Var1   :   W5N
Var2   :   W4N
#Sampl:    -1
```

The sequence W5N is shown in Fig. 4-21.

4-7 EXERCISES

1. Refer to Ex. 4-2. Pad the sequence XNL with 64 additional zeros to extend its length to 128.

a. Compute the FFT of XNL. Plot the magnitude and the phase of the FFT. Compare the results with those of Ex. 4-2. Does zero-padding change the general shape of the frequency spectrum?

b. Compare even-numbered samples of the magnitude just computed with the magnitude obtained in Ex. 4-2, e.g., compare $X_{new}(6)$ with $X_{old}(3)$, and $X_{new}(14)$ with $X_{old}(7)$, ..., and so on.

2. Consider the following two sequences:

$$x_1(n) = \{3 \;\; 4.2 \;\; 11 \;\; 0 \;\; 7 \;\; -1 \;\; 0 \;\; 2\}, \qquad n = 0, \ldots, 7$$

$$x_2(n) = \{1.2 \;\; 3 \;\; 0 \;\; -0.5 \;\; 2\}, \qquad n = 0, \ldots, 4$$

a. Compute the convolution of $x_1(n)$ and $x_2(n)$ using the *Convolve* function. What is the length of the result?

b. In some cases, it may be computationally more efficient to use Fourier transforms to compute the convolution of two sequences. First, we will attempt to do it in a way that a not-so-careful student might use. Extend the sequence $x_2(n)$ to nine points by zero-padding so that both sequences have the same length. Compute nine point DFTs $X_1(k)$ and $X_2(k)$.

$$X_1(k) = \text{DFT}\{x_1(n)\} \qquad X_2(k) = \text{DFT}\{x_2(n)\}$$

Multiply $X_1(k)$ and $X_2(k)$; then compute the inverse DFT of the product. Take the real part to remove the imaginary part, which is due to round-off error. How does this result compare with the correct linear convolution sequence obtained in part (a)? How many samples are correct? Why?

c. What is the minimum DFT length that should be used if we are to obtain the true linear convolution of $x_1(n)$ and $x_2(n)$? Zero-pad both $x_1(n)$ and $x_2(n)$ to extend their lengths to the minimum required DFT length. Repeat part (b).

d. Pad $x_1(n)$ and $x_2(n)$ with five more zeros, so that their lengths exceed the minimum DFT length required for linear convolution by five samples. Repeat part (c). How does the excess length affect the result?

3. Two linear systems are characterized by their impulse responses $h_1(n)$ and $h_2(n)$.

$$h_1(n) = \{2 \;\; 3.2 \;\; 1 \;\; -0.5 \;\; 1.2 \;\; 4\}, \qquad n = 0, \ldots, 5$$

$$h_2(n) = \{3 \;\; -1.5 \;\; 0 \;\; 2.6\}, \qquad n = 0, \ldots, 3$$

a. These systems are to be connected in cascade, as follows:

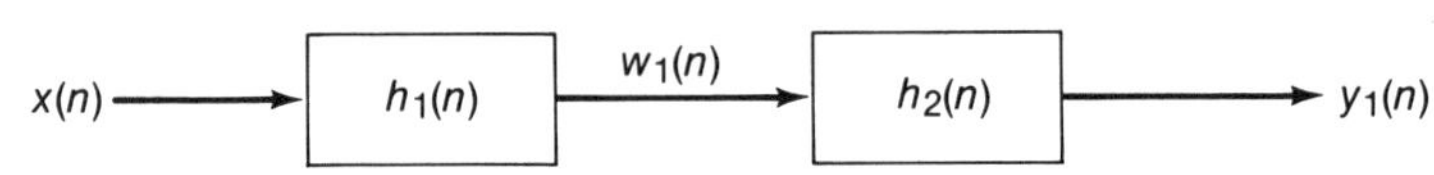

Figure p4-3a

Generate an arbitrary input sequence $x(n)$. Compute the output sequence $y_1(n)$ as dictated by the block diagram; i.e., first convolve $x(n)$ and $h_1(n)$ to obtain $w_1(n)$, then convolve $w_1(n)$ and $h_2(n)$ to obtain $y_1(n)$.

b. The order of cascading two systems is to be changed as shown. Repeat part (a). Compare the output sequences $y_1(n)$ and $y_2(n)$.

4. The second block (i.e., $h_2(n)$) in Exercise 3 is replaced with a nonlinear system with input-output relationship

$$y(n) = 0.01[x(n)]^2$$

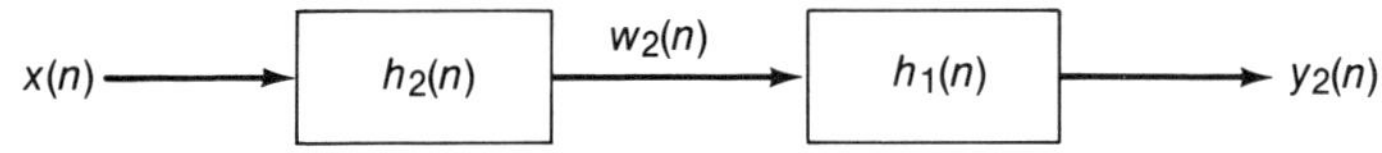

Figure p4-3b

The first block remains the same. Repeat Exercise 3. How do the output sequences compare now?

5. Consider the five-sample sequence $x(n)$ given as

$$x(n) = \{3\ 2\ 7\ 1\ 4\}$$

a. Compute the five-point DFT of $x(n)$. Multiply the DFT by the complex exponential sequence $\exp(-j2\pi k/5)$ and compute the inverse DFT of the product, i.e.,

$$X(k) = \text{DFT}\{x(n)\}$$

$$x_1(n) = \text{INV} - \text{DFT}\{X(k).\exp(-j2\pi k/5)\}$$

Take the real part of $x_1(n)$ to get rid of the imaginary part which is due to round-off errors. Compare $x_1(n)$ with $x(n)$. Is it a circularly shifted version of $x(n)$?

b. What would you change in part (a) to obtain a circular shift by three samples?

c. How would you modify the technique so that linear shifts can be obtained?

6. Refer to Exercise 4 of Chapter 3.

a. Compute the FFT of the suppressed carrier double sideband signal $x_{SC}(n)$. Plot the magnitude of the FFT.

b. The signal is to be demodulated using synchronous demodulation as follows

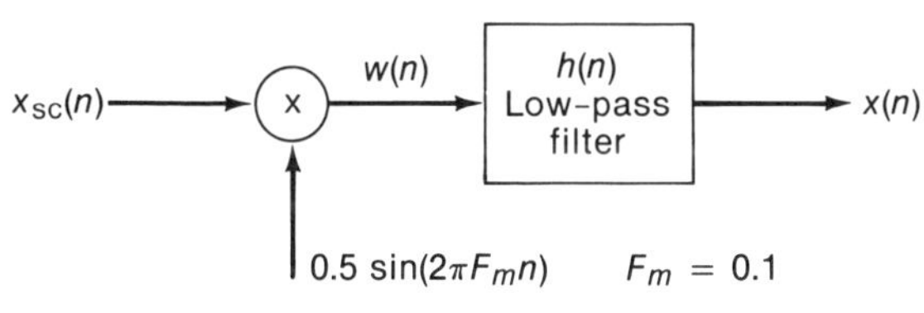

Figure p4-6

Compute the FFT of $w(n)$. Plot its magnitude and phase.

c. The impulse response $h(n)$ of a suitable low-pass filter is given in the data file EXER4-6.DSP. Use it with the *Convolve* function to obtain $x(n)$.

d. Repeat part (c), but change the phase of the sine term by 45°, i.e., use

$$\sin\left(2\pi F_m n + \frac{\pi}{4}\right)$$

With this change, we will be simulating a case where the local oscillator signal is not in perfect synchronization with the original carrier. How does this affect the demodulated signal $x(n)$?

e. Repeat part (c) with an incorrect local oscillator frequency, $F_m = 0.11$. Replace the sinusoid with

$$\sin[2\pi(0.11)n]$$

How is $x(n)$ affected by the mismatch of two frequencies?

7. Generate a 32-point complex sequence $x(n)$ and use it to numerically verify the following theorem:

Theorem If $x(n)$ and $X(k)$ are a discrete Fourier transform pair, then the DFT of the real part of $x(n)$ is equal to the conjugate symmetric part of $X(k)$.

Hint: Compute conjugate symmetric part of $X(k)$ using the *Even* function. Take the real part of $x(n)$ and compute its DFT. Compare.

8. Using the *inv-Z* function, generate the sequence

$$x(n) = (0.9)^n \sin(0.2n), \qquad n = 0, ..., 99$$

Hint: First, find the rational z-transform of $x(n)$ either by hand or from a z-transform table. Then enter the numerator and the denominator using the keyboard entry method. Use *inv-Z* to get as many samples as needed.

9. Consider a linear system described by the following z-domain transfer function:

$$H(z) = \frac{z-1}{(z-0.25)(z-0.5)}$$

a) Use the *inv-Z* function to compute the first 100-samples of the unit-step response of this system.
b) Express the system function in parallel form as

$$H(z) = H_1(z) + H_2(z)$$

Determine unit-step responses of parallel blocks $H_1(z)$ and $H_2(z)$, and add the results. Compare to the response found in part (a).

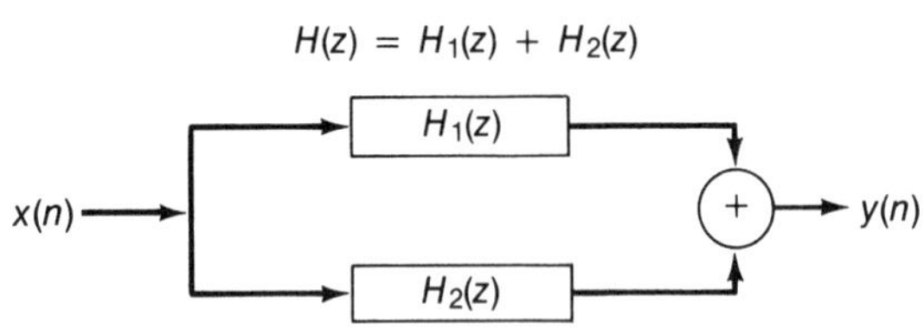

Figure p4-9

10. Consider a linear system with the z-domain transfer function

$$H(z) = \frac{1}{1 - z^{-N}}$$

a. For $N = 10$, compute the first 150 samples of the impulse response of this system.

b. Repeat part (a) for $N = 20$. Compare with the impulse response found in part (a).

c. How can this system be used for generating a periodic sequence from a finite-length input sequence?

11. In this exercise, we look at the aliasing phenomenon. To begin with, generate and plot 300 samples of the sequence

$$x(n) = \sin[2\pi(0.1)n], \qquad n = 0, \ldots, 299$$

a. Decimate $x(n)$ with a factor of 2 to obtain a new sequence $x_1(n)$. Plot $x_1(n)$. Does it look like $x(n)$?

b. Try to reconstruct $x(n)$ (approximately) from $x_1(n)$ by first inserting zeros between samples and then filtering it through a FIR low-pass filter. Is it possible? (*Note:* The impulse response of a FIR low-pass filter with normalized cutoff frequency 0.2 is given in the data file EXER4-11.DSP)

c. Repeat parts (a) and (b) for a decimation factor of 6. Do your answers to the above questions change?

12. The sampling rate of a signal is to be changed by a non-integer factor $\frac{7}{5}$. To accomplish this, we can use a decimator and an interpolator in cascade.

a. Use the sequence $x(n)$ in Exercise 11. First decimate by a factor of 5 and then interpolate by a factor of 7. Plot the resulting sequence.

b. Repeat part (a), but this time interpolate first and decimate later. Plot the resulting sequence. Is it different from the sequence of part (a)? Why?

13. A signal with an echo can be represented in the form

$$x(n) = w(n) + Aw(n - m)$$

where the first term, $w(n)$, is the original signal, and the second term, $Aw(n - m)$, is the echo.

a. Using white noise with zero mean and unit variance for $w(n)$, generate 300 samples of a sequence $x(n)$. Use $A = 0.4$ and $m = 17$.

b. Compute the autocorrelation estimate of $x(n)$. Observe the peak at the location of the echo. Can the autocorrelation function be used to detect the presence of an echo in a random signal?

c. Repeat (a) and (b) with $A = 0.5, m = 32$.

5

PC-DSP Graphics

In the examples of previous chapters, we have relied on the quick-plot feature to obtain screen graphics. This is the easiest method for graphing variables in PC-DSP. Pressing the [F2] key produces a screen plot of the most recently accessed plottable sequence. Graphics parameters such as the title, plotting mode (either continuous or discrete), axis limits, axis labels, and data labels are selected by the program using a built-in set of rules. In some cases, it may be desirable to change some of these parameters. The *Graphics* component of PC-DSP allows the user to customize screen plots. It is also possible to overlay two plots.

5-1 SPECIFYING VARIABLES TO BE GRAPHED

The variables to be graphed are specified by selecting the *Variables* option of the *Graphics* menu and answering the prompts. PC-DSP recognizes three graph variables, one for the x-axis and two for the y-axis. All graph variables must be real. Of the three variables, only the first y-axis variable is required, and the other two are optional. If the x-axis variable is not specified, PC-DSP internally generates a temporary integer sequence starting with 0 for use as the x-axis variable. The second y-axis variable is used for overlaying two plots. If more than one graph variable is specified, then all specified variables must have the same number of samples. We will proceed with a simple example.

Example 5-1:

In this example, we will generate a sinusoidal sequence and plot it using the *Graphics* component. Unless you have just started a PC-DSP session, use the *Reset* function of the *Graphics* menu to ensure that all graphics parameters are set to their initial values. Now generate a 101-point sequence YN using the *Generate / Sine* function of the *Data* menu with the following parameter values:

$A = 1.0, B = 0.1, C = 0, \#\text{Sampl} = 101.$

Select *Variables*, and make the following entries:

```
X-Var  :  ________

Y-Var1 :   YN
          ________

Y-Var2 :
          ________
```

Note that the <ENTER> key is pressed without typing anything in response to the first and the last prompts. Now select *Display* from the *Graphics* menu. The sequence YN is plotted in discrete mode (see Fig. 5-1).

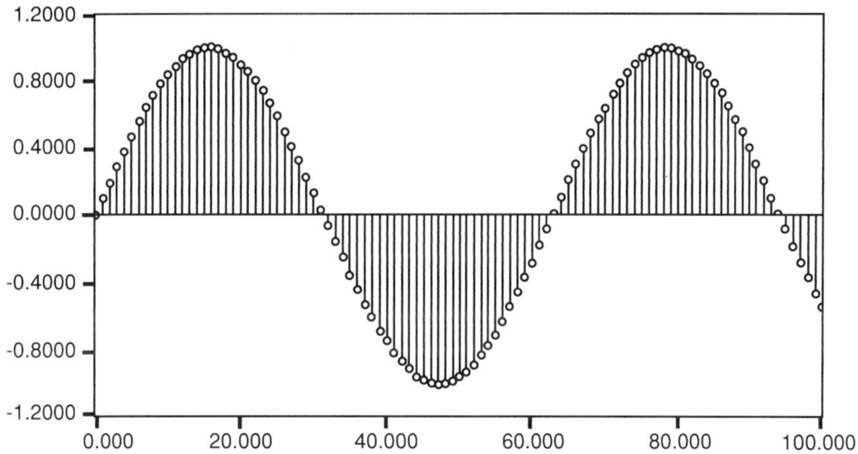

Figure 5-1. Sequence YN of Ex. 5-1 plotted with default graph settings.

Example 5-2:

In the plot of Ex. 5-1, no x-axis variable was specified. Therefore a temporary integer sequence was generated by PC-DSP for use as the x-axis variable. Now assume that the sequence YN was obtained by sampling a continuous-time sinusoid at 0.01-second intervals. Generate the x-axis variable TIME using the *Ramp* function of the *Data* menu with the following parameter values:

$\text{Init} = 0, \text{Incr} = 0.01, \#\text{Sampl} = 101$

Now select *Graphics / Variables*, and enter the graph variables as follows:

```
X-Var  :  TIME
Y-Var1 :  YN
Y-Var2 :
```

Select the *Display* function to obtain the plot shown in Fig. 5-2.

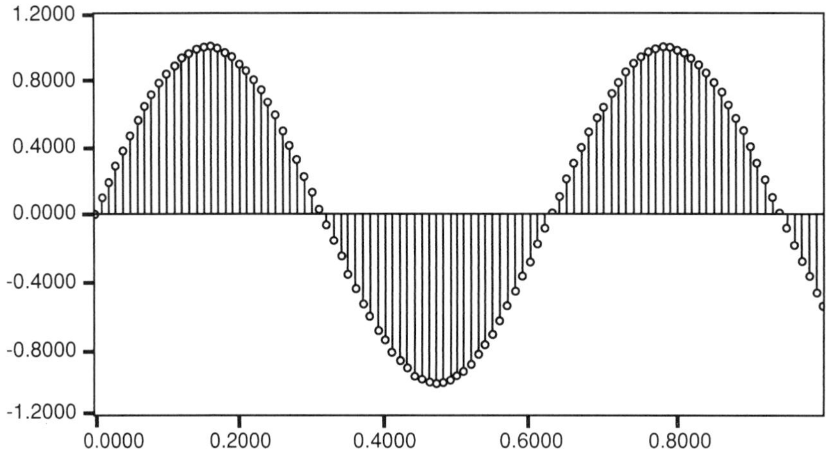

Figure 5-2. Plot of YN versus TIME.

Example 5-3:

In this example, we will generate another 101-point sequence and overlay it with YN of Ex. 5-2. Generate a sequence WN using the *Generate / Cosine* function of the *Data* menu with the following parameter values:

$$A = 0.8, B = 0.1, C = 0, \#\text{Sampl} = 101.$$

Select *Variables*, and make the following entries:

```
X-Var  :  TIME
Y-Var1 :  YN
Y-Var2 :  WN
```

When finished, select *Display* to obtain the screen plot (Fig. 5-3). The second y-axis variable is plotted as a continuous curve with a dashed line. It is difficult to make sense out of two overlaid discrete plots; that is why the second plot is continuous even though the plotting mode is discrete.

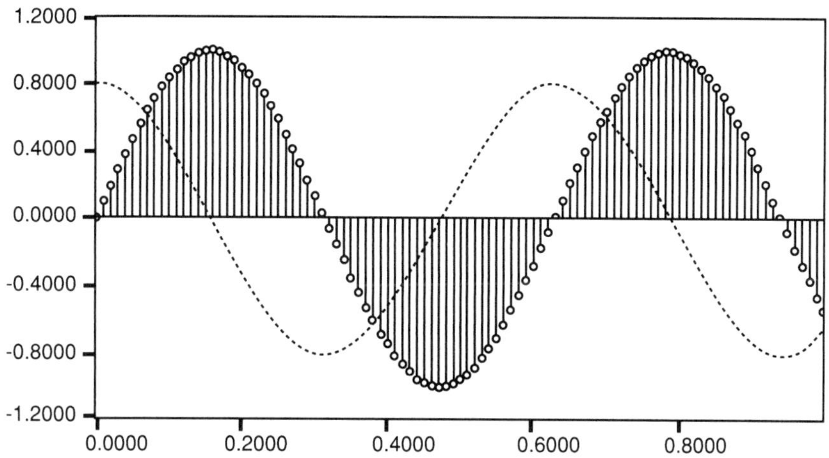

Figure 5-3. Overlay of YN versus TIME and WN versus TIME.

5-2 GRAPHICS FORMATTING OPTIONS

Graph settings can be changed by selecting *Options* from the *Graphics* menu and answering program prompts. Most prompts are self-explanatory.

Example 5-4:

Modify the plot of Ex. 5-3 so that both variables are plotted in continuous mode for the range of TIME from 0.0 to 0.2 seconds. Title the graph "Sinusoids." Use "Time (sec)" for the x-axis label and "Amplitude" for the y-axis label. For the x-axis, place data labels every 0.04 seconds, and use three digits after the decimal point. The y-axis should extend from -1.5 to 1.5 with data labels placed at increments of 0.5. Y-axis data labels should have two digits after the decimal point. Also display a grid.

Select *Options* from the *Graphics* menu. The first prompt of the program is about the plotting mode.

```
Mode     :    D
```

Two choices are listed in an adjacent help box as

```
D = Discrete
C = Continuous
```

and the current mode is indicated with a D on the answer line. Enter a C to change the mode to continuous. Then answer the next three questions as follows:

```
Title    :    Sinusoids

X-Label  :    Time (sec)
```

```
Y-Label :      Amplitude
```

Next, specify axis limits with the following entries:

```
X-min   :   0.0          X-max   :   0.2
Y-min   :   -1.5         Y-max   :   1.5
```

Desired data label increments are 0.04 for x-axis data labels and 0.5 for y-axis data labels. Data label formats (number of digits after decimal point) are 3 and 2, respectively. Make the following entries:

```
Data label incr.    --->  X-axis :   0.04     Y-axis :   0.5
Data label format   --->  X-axis :   3        Y-axis :   2
Grid ?  (Y/N)  :   Y
```

Now select *Display* to obtain the plot shown in Fig. 5-4. Do not press the [F2] key. Pressing the [F2] key generates a quick-plot with default graph settings. When graph parameters are modified, the *Display* function must be used.

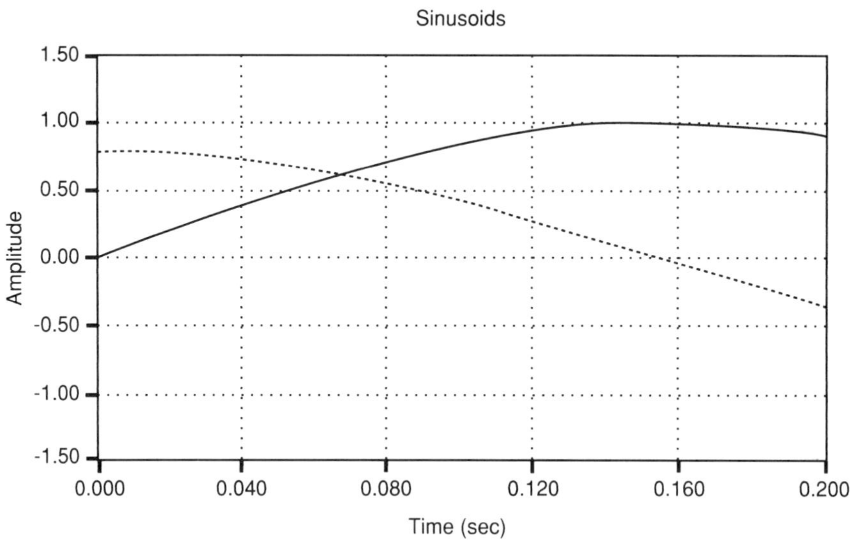

Figure 5-4. Screen plot obtained in Ex. 5-4.

Example 5-5:

Plot the magnitude of the AM frequency spectrum computed in Ex. 4-3. This time, use radian frequencies for the x-axis, and plot the magnitude in continuous mode.

The sequence XAMKL can be loaded from the data file EX4-3.DSP. Use the *Ramp* function with the following parameter values to generate a variable RFREQ that contains radian frequencies:

$$\text{Init} = 0, \text{Incr} = 2\pi/1024, \#\text{Sampl} = 512$$

Select *Options* to set the mode, title, axis labels, data labels, and grid. The resulting plot is shown in Fig. 5-5.

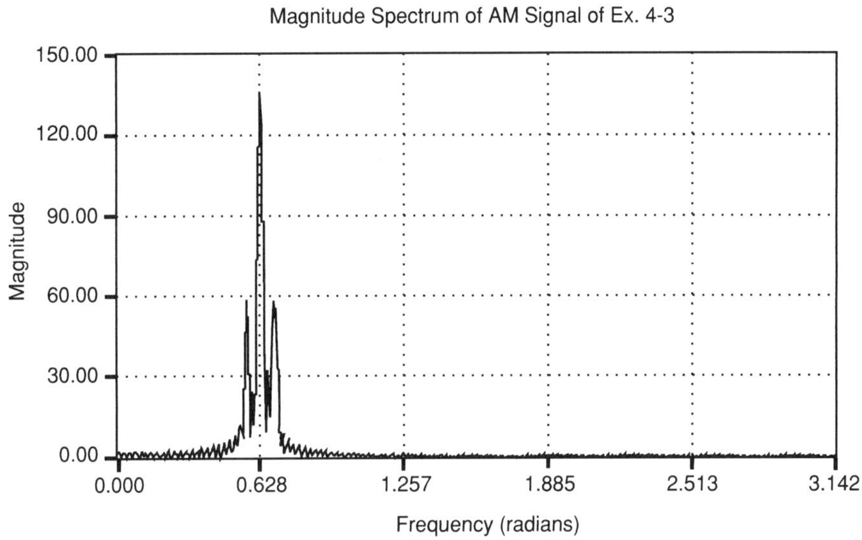

Figure 5-5. Magnitude spectrum for AM signal.

5-3 LOADING AND SAVING GRAPHICS FILES

Load and *Save* functions of the *Graphics* menu are used for writing and reading graphics data files. This capability permits the user to save plots for later viewing or printing. Graphics files are in binary format similar to the DSP files explained in Chapter 3. The default extension for graphics data files is GRP. A graph must be displayed on the screen before it can be saved into a file.

Example 5-6:

A sample graphics file with the name MYGRAPH.GRP is provided in the Utilities and Examples Disk. To load the file, select *Load* from the *Graphics* menu, and enter the file name.

```
Filename [.GRP] :   MYGRAPH
```

When the file is loaded, PC-DSP displays the message

```
Loaded graph file :  MYGRAPH.GRP
```

At this point, the graph can be displayed by selecting *Display* from the *Graphics* menu.

6

Finite Impulse Response Filters

In this chapter, design and analysis of finite impulse response (FIR) filters using PC-DSP are discussed. We first look at the analysis options for FIR filters and then cover in detail the three FIR filter design techniques that are available.

6-1 ANALYSIS OF FIR FILTERS

In PC-DSP, the description of a FIR filter consists of the impulse response of the filter stored as a named sequence. Our first example will involve a very simple four-tap FIR filter.

Example 6-1:

Compute and plot the magnitude, the log-magnitude, and the phase of the system function $H(e^{j\omega})$ for a four-sample averaging filter that is described by the difference equation

$$y(n) = \frac{1}{4}[x(n) + x(n-1) + x(n-2) + x(n-3)]$$

For any value of n, the output sample $y(n)$ is computed as the arithmetic average of the most recent four input samples. It can be shown that the impulse response of this system has four nonzero samples, i.e.,

$$h(n) = \{\frac{1}{4} \quad \frac{1}{4} \quad \frac{1}{4} \quad \frac{1}{4}\}, \qquad n = 0, 1, 2, 3$$

We will generate the four-sample impulse response HN using the keyboard entry method. Select the *Read* function of the *System* menu, and enter the following:

```
Variable name :     HN

Sample values :     1/4 ; 1/4 ; 1/4 ; 1/4
```

Now select *Filter*, then select *FIR-filters*, and finally *Analyze*. A data entry window will be opened in the lower half of the screen. The impulse response description of the FIR filter to be analyzed is in the variable HN, so answer the first program prompt as

```
Variable name for filter data :     HN
```

The next prompt is about the type of analysis to be performed.

```
Analysis type :     M
```

Possible choices are listed in a help box below the prompt.

```
M = Magnitude plot
L = Log-magnitude plot
P = Phase plot
```

The default analysis type is M. Since we want to obtain a magnitude plot first, press the <ENTER> key to accept this default.

Now we need to specify the name of a variable that will hold the data points for the desired magnitude plot. PC-DSP computes 256 points of the magnitude response equally spaced in frequency from 0 to π radians. (512 points from 0 to 2π radians are computed for FIR filters with complex impulse response.) These data points are stored in a named sequence so that they can be graphed and/or tabulated. The default name is TEMPR.

```
Variable name for analysis results :     TEMPR
```

Press the <ENTER> key to accept the default variable name. After a few seconds, the following two lines are displayed:

```
Variable = TEMPR   , Type = R , Size =  256
Variable = NFREQ   , Type = R , Size =  256
```

Two variables are generated for a magnitude plot. The first variable, TEMPR, holds magnitude values, as explained earlier. The second variable, NFREQ, contains 256 normalized frequency values that correspond to the magnitude points in TEMPR. The magnitude plot can now be obtained by pressing the [F2] function key.

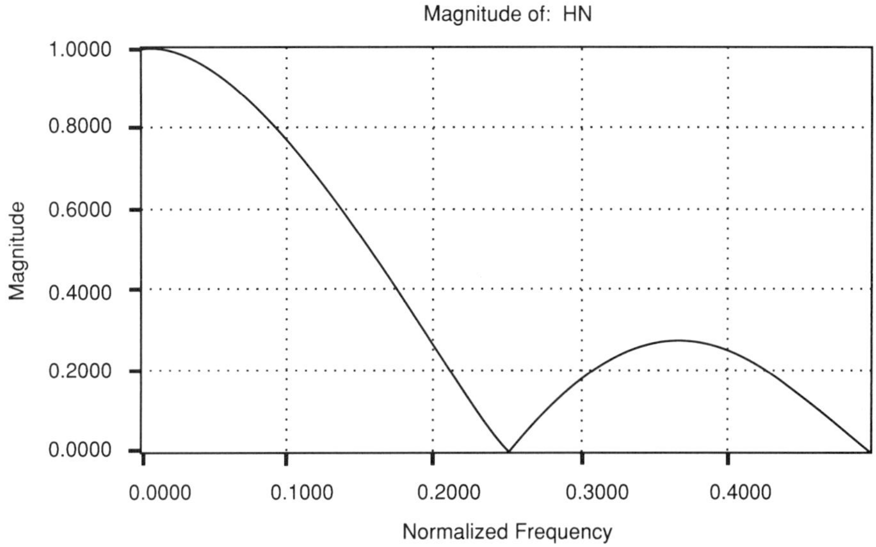

Figure 6-1. Magnitude plot for the four-tap averaging filter of Ex. 6-1.

The magnitude is graphed as a function of the normalized frequency variable F, which is related to the radian frequency variable ω by

$$\omega = 2\pi F \tag{6-1}$$

Log-magnitude and phase plots can be obtained by selecting the *Analyze* function again and answering the analysis type prompt with L and P, respectively (see Figs. 6-2 and 6-3).

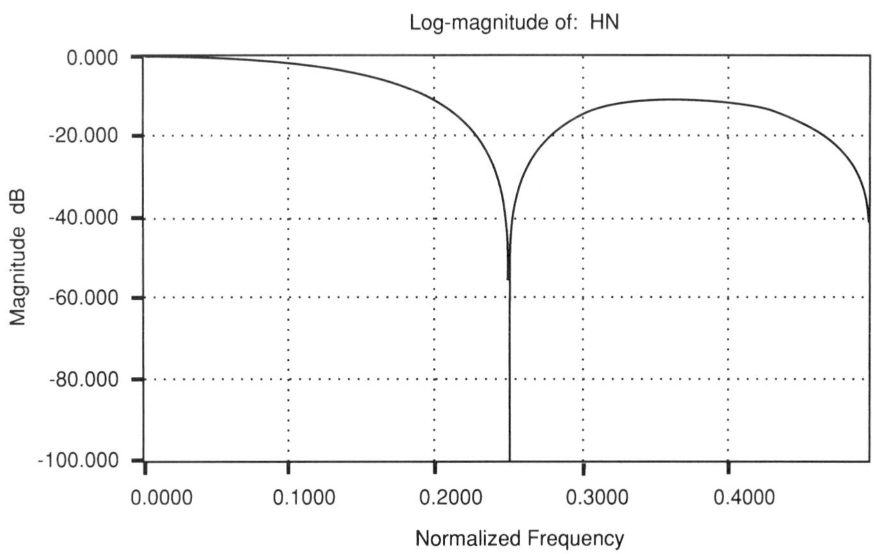

Figure 6-2. Log-magnitude plot for the four-tap averaging filter of Ex. 6-1.

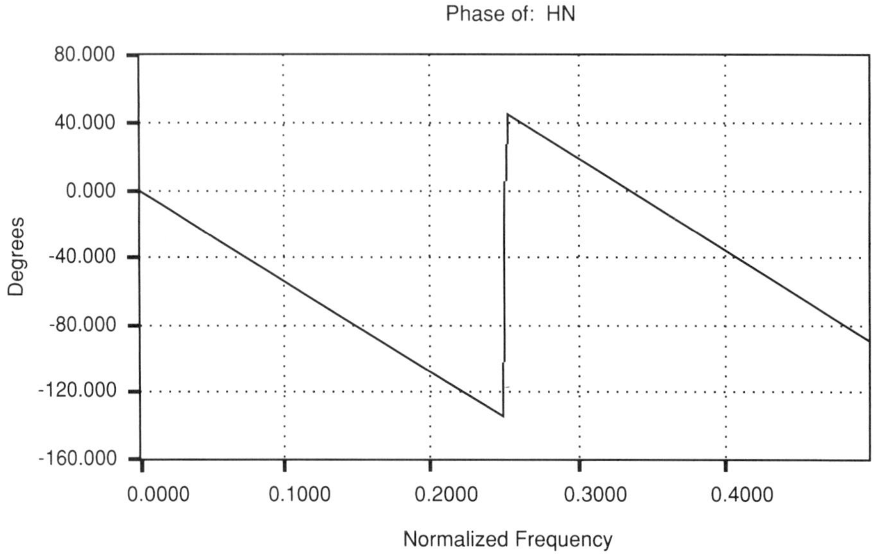

Figure 6-3. Phase plot for the four-tap averaging filter of Ex. 6-1.

The impulse response of a practical FIR filter would typically be real-valued and less than 100 samples long. Most FIR filters used in practice also have linear phase, since this is the main advantage of FIR filters over IIR filters. In the following sections of this chapter, we will see that FIR filters designed by PC-DSP are also linear-phase filters limited to 100 real coefficients. Nevertheless, for analysis purposes, it is possible to produce magnitude, log-magnitude, and phase plots for FIR filters that may be complex and up to 512 samples long with nonlinear phase.

Example 6.2:

Most digital signal processing textbooks give a graphical comparison of various discrete window functions in terms of the log-magnitude of the Fourier transform. In this example, we will generate the log-magnitude for the Blackman window of Ex. 3-13. If you no longer have the 50-point Blackman window sequence BLWIN in memory, it can be loaded from the data file EX3-13.DSP. Select *FIR-filters / Analyze*, and proceed as in Ex. 6-1. The log-magnitude is shown in Fig. 6-4. By tabulating the log-magnitude response with the Print function of the System component, it can be verified that the half-width of the main lobe is 0.0625 (normalized frequency) or, equivalently, 0.125π radians. Minimum side-lobe attenuation is 58.7 decibels.

6-2 FOURIER SERIES DESIGN OF FIR FILTERS

The simplest method of designing a linear-phase FIR filter is the Fourier series method. Design specifications are given in the form of an ideal filter to be approximated. The impulse response of this ideal filter is computed using a discrete-time inverse Fourier

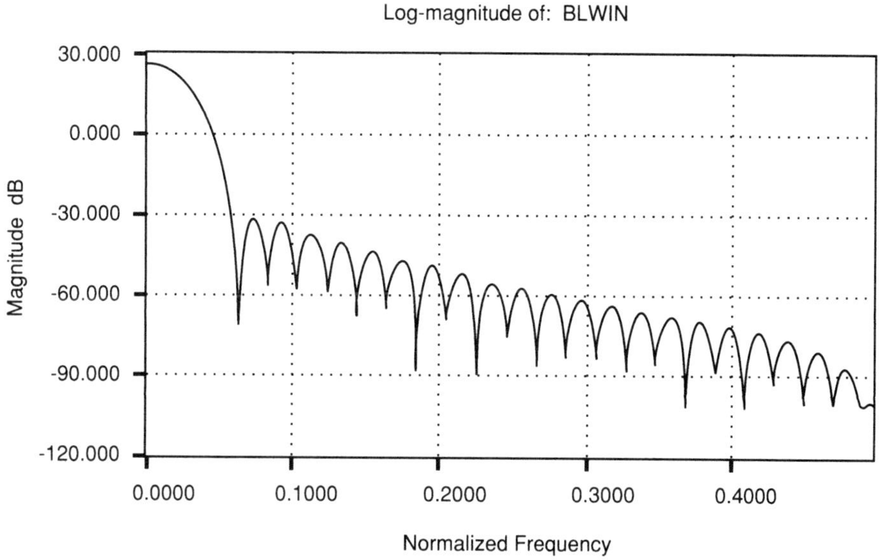

Figure 6-4. Log-magnitude for a 50-sample Blackman window.

transform. Since this impulse response is of infinite length, it is multiplied by a symmetric window function, so that a finite-length sequence is obtained while retaining the linear-phase property. We will illustrate the Fourier series design of FIR filters with two examples.

Example 6-3:

Using the Fourier series method, design a 51-tap linear-phase low-pass filter to approximate the ideal filter characteristic shown in Fig. 6-5. Use a rectangular window.

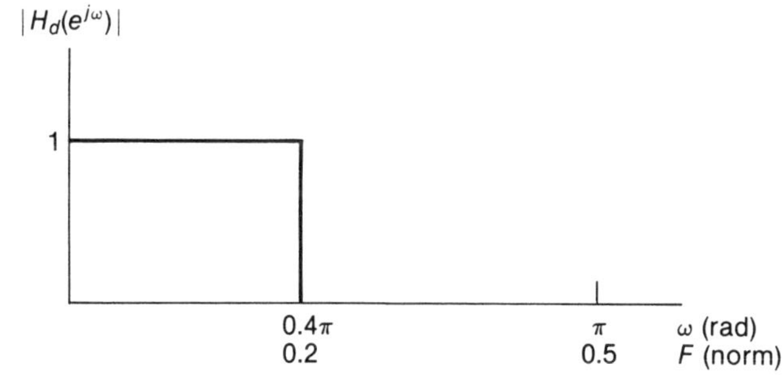

Figure 6-5 Specifications for the filter of Ex. 6-3.

Begin by selecting the *Design* function from the *FIR-filters* submenu. The first program prompt to be answered is about the design method to be used.

```
Design method :  F

F = Fourier series design
P = Parks - McClellan algorithm
L = Least squares design
```

The Fourier series design method is the default choice. Press the <ENTER> key to accept the default. In the next step, we need to specify the filter type desired. Six different types of filters can be designed using the Fourier series method.

```
Filter type :  L
              ___
                   L = Low-pass
                   H = High-pass
                   B = Band-pass
                   R = Band-reject
                   D = Differentiator
                   T = Hilbert trans.
```

We will accept the default answer *L* for a low-pass filter. Answer the next two program prompts as follows:

```
Filter order :     51
                  ____

Window type :  R
              ___
                   R = Rectangular
                   T = Triangular
                   B = Blackman
                   H = Hamming
                   A = hAnning
```

Now, we need to specify the corner frequency for the low-pass filter. In PC-DSP, frequencies are always specified in terms of normalized values.

```
F1  :     0.2
        _______
```

Thus, filter specifications are complete. If you have a printer connected to the computer and wish to obtain a hard copy of the design procedure, answer the prompt

```
Printer report ?  :    Y
                      ___
```

with a *Y* for yes. To skip the printer report, answer with an *N*. The program will take several seconds to design the filter and then request a variable name to store the impulse response of the designed filter. Let's use the name FIR1.

```
Variable name for impulse response :    FIR1
                                       ______
```

This should generate the familiar response

```
Variable = FIR1     , Type = R , Size =   51
```

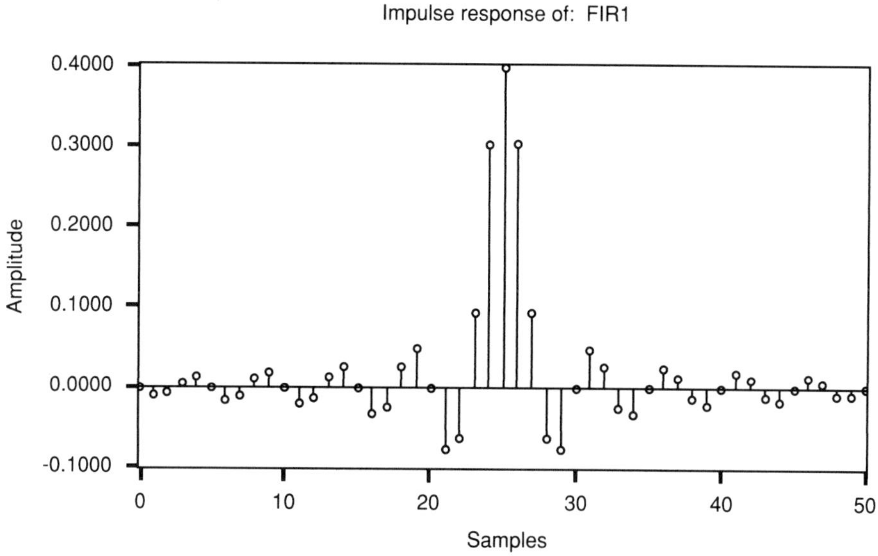

Figure 6-6. Impulse response of the filter FIR1.

```
===== PC-DSP    FIR Filter Design Repot =====

   Design method : Fourier series
   Filter type   : Lowpass
   Window type   : Rectangula
   Filter oder   :  51

   Nomalized corner frequencies:

    F1 = 0.20000

   Impulse response:

     H( 0) =  5.230489E-08    H(50) =  5.230489E-08
     H( 1) = -1.261376E-02    H(49) = -1.261376E-02
     H( 2) = -8.134723E-03    H(48) = -8.134723E-03
     H( 3) =  8.504418E-03    H(47) =  8.504418E-03
     H( 4) =  1.441576E-02    H(46) =  1.441576E-02
     H( 5) =  4.744787E-08    H(45) =  4.744787E-08
     H( 6) = -1.593318E-02    H(44) = -1.593318E-02
     H( 7) = -1.039437E-02    H(43) = -1.039437E-02
     H( 8) =  1.100571E-02    H(42) =  1.100571E-02
     H( 9) =  1.892068E-02    H(41) =  1.892068E-02
     H(10) =  3.935283E-08    H(40) =  3.935283E-08
     H(11) = -2.162361E-02    H(39) = -2.162361E-02
     H(12) = -1.439219E-02    H(38) = -1.439219E-02
     H(13) =  1.559145E-02    H(37) =  1.559145E-02
     H(14) =  2.752098E-02    H(36) =  2.752098E-02
     H(15) =  4.744787E-08    H(35) =  4.744787E-08
     H(16) = -3.363672E-02    H(34) = -3.363672E-02
     H(17) = -2.338727E-02    H(33) = -2.338727E-02
     H(18) =  2.672824E-02    H(32) =  2.672824E-02
     H(19) =  5.045512E-02    H(31) =  5.045512E-02
     H(20) =  4.744787E-08    H(30) =  4.744787E-02
     H(21) = -7.568265E-02    H(29) = -7.568265E-02
     H(22) = -6.236598E-02    H(28) = -6.236598E-02
     H(23) =  9.354888E-02    H(27) =  9.354888E-02
     H(24) =  3.027307E-01    H(26) =  3.027307E-01
     H(25) =  4.000000E-01

 Results stored in the variable  :  FIR1
```

Figure 6-7. Printer report for FIR filter design.

and a quick-plot of the impulse response can be obtained by pressing the [F2] key (see Fig. 6-6).

The designed filter, FIR1, can be analyzed using the *Analyze* function of the *FIR-filters* submenu, as explained in the previous section. The magnitude plot is shown in Fig. 6-8. Note the oscillatory behavior in the vicinity of the cutoff frequency. This is the well-known Gibbs phenomenon due to the rectangular window used.

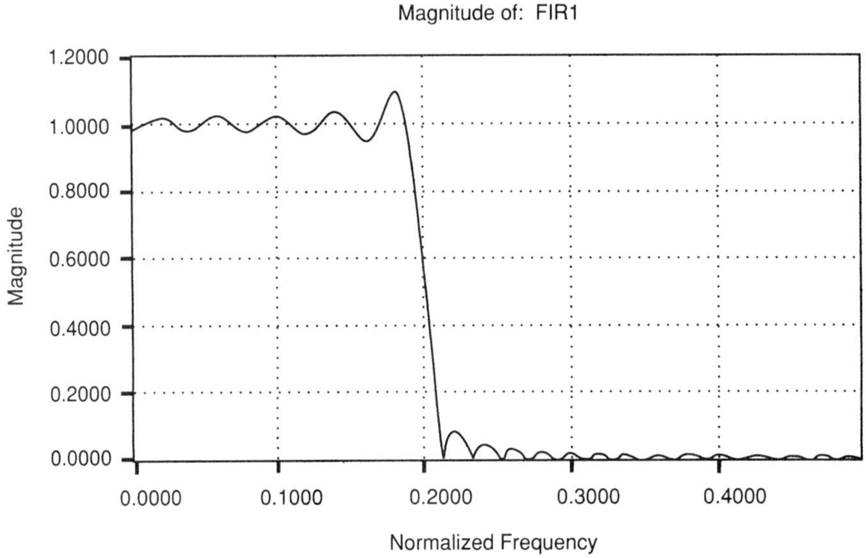

Figure 6-8. Magnitude response of the filter FIR1.

Example 6-4:

Using the Fourier series method, design a 40-tap linear-phase bandpass filter to approximate the ideal filter characteristic shown in Fig. 6-9. Use a Hamming window.

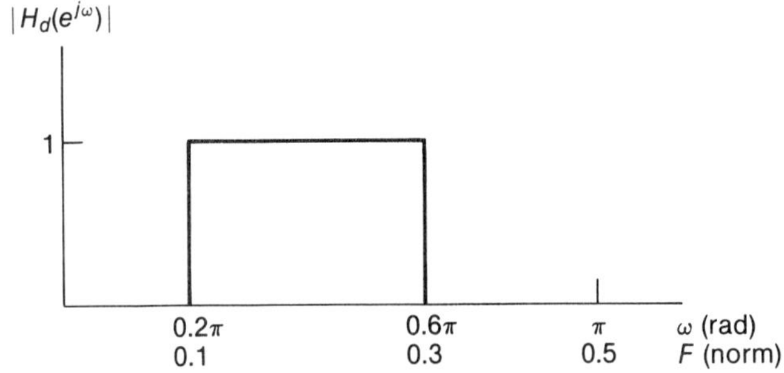

Figure 6-9. Specifications for the filter of Ex. 6-4.

Select the *Design* function from the *FIR-filters* submenu again, and enter the following parameter values:

```
Design method :   F          Filter type :   B
```

```
Filter order  :   40          Window type :   H
F1   :   0.1                  F2   :   0.3
```

Two corner frequencies, F_1 and F_2, are specified for a band-pass filter. Impulse response and magnitude plots are given in Figs. 6-10 and 6-11. Note that the use of a Hamming window virtually eliminates the Gibbs oscillations; however, sharpness of the transition from one band to another is greatly reduced.

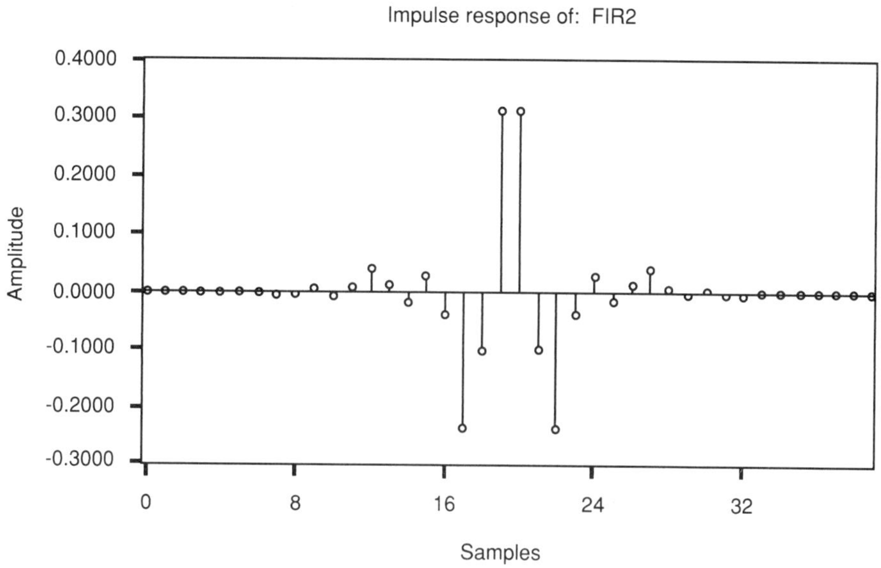

Figure 6-10. Impulse response of the filter designed in Ex. 6-4.

Example 6-5:

Using the Fourier series method with a Blackman window, design a 40-tap differentiator to approximate the desired magnitude response

$$|H_d(e^{j\omega})| = 2F$$

shown graphically in Fig. 6-12.

Select *FIR-filters / Design*, and enter the following parameter values:

```
Design method :   F            Filter type :   D
Filter order  :   40           Window type :   B
Slope  :   2
Variable name for impulse response :   FIR3
```

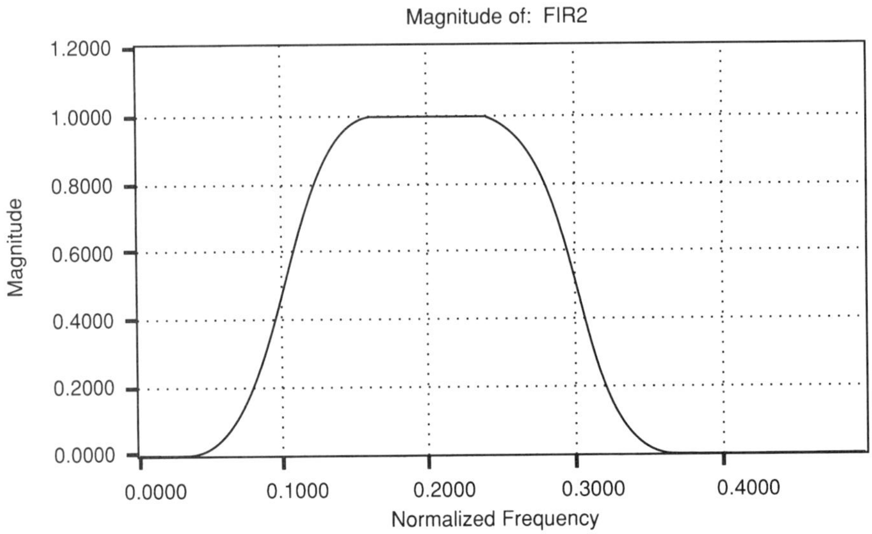

Figure 6-11. Magnitude of the filter designed in Ex. 6-4.

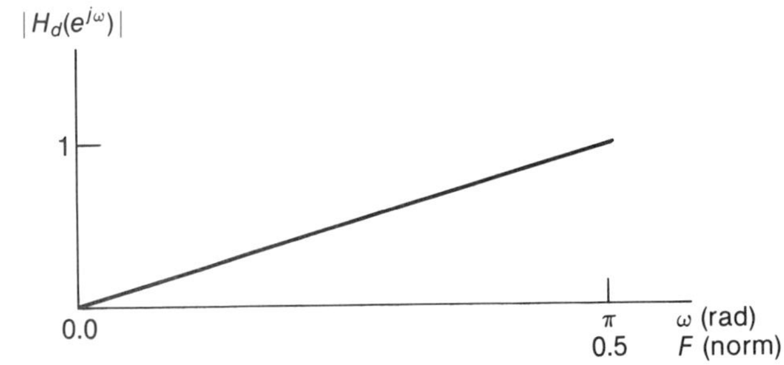

Figure 6-12. Specifications for the filter in Ex. 6-5.

6-3 PARKS-MCCLELLAN ALGORITHM

The Fourier series method of designing FIR filters is quite simple and fast. Nevertheless, the Gibbs phenomenon caused by the truncation of an infinite-length impulse response may be a drawback in some applications. Although a window function can be used to eliminate Gibbs oscillations, it then becomes difficult to have precise control over the cutoff frequency.

Using the design algorithm due to Parks and McClellan, it is possible to design FIR filters that are optimum in approximating a set of magnitude specifications with minimum peak error. The desired filter magnitude is specified for a set of frequency bands that are disjoint; i.e., each band is separated from its neighbors with transition bands. A weight factor can also be assigned to each frequency band reflecting the relative importance of that band compared to others. Minimax error criterion is used so that the maximum value of the approximation error is minimized, resulting in an equiripple filter.

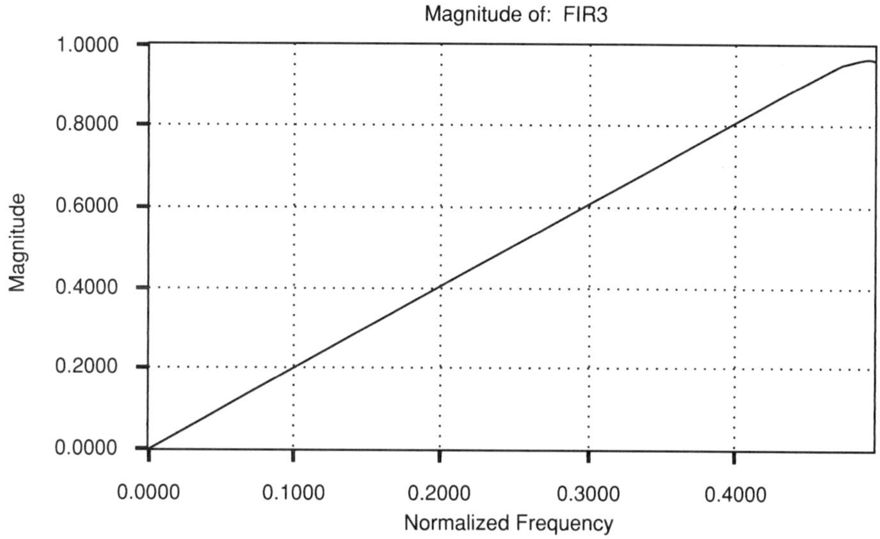

Figure 6-13. Magnitude of the filter designed in Ex. 6-5.

Example 6-6:

Design a 55-tap equiripple FIR band-pass filter to approximate the desired magnitude response

$$|H_d(e^{j\omega})| = \begin{cases} 0, & 0 < \omega < 0.2\pi & 0 < F < 0.1 \\ 1, & 0.25\pi < \omega < 0.45\pi & 0.125 < F < 0.225 \\ 0, & 0.5\pi < \omega < \pi & 0.25 < F < 0.5 \end{cases}$$

Use uniform weighting for all frequency bands.

Begin by selecting the *Design* function of the *FIR-filters* submenu. When prompted for the design method, type *P* for the Parks-McClellan algorithm. The following prompt will be displayed:

```
Filter type :    M
```

Possible answers are again listed in a help box.

```
M = Multiband filter
D = Differentiator
T = Hilbert trans.
```

Accept the default answer M by pressing the <ENTER> key. The next item to be entered is the filter order.

```
Filter order :    55
```

Next, enter the number of frequency bands:

```
# of Bands   :     3
```

The filter we wish to design has three disjoint frequency bands. Now we need to enter the specifications for each band. For a multiband filter, PC-DSP requires the lower edge of the first band to be equal to 0.0 and the upper edge of the last band to be equal to 0.5, both in normalized frequencies. Enter the following parameter values:

	Band 1	Band 2	Band 3
Lower Edge	0.0	0.125	0.25
Upper Edge	0.1	0.225	0.5
Value	0	1	0
Weighting	1	1	1

The next prompt of the program is about the printer report option. After the printer report question is answered, PC-DSP will start working on the filter design problem. The Parks-McClellan algorithm takes somewhat longer than the Fourier series method, due to its iterative nature. When the design is completed, the program will ask for the name of the variable to hold the impulse response of the designed filter. We will call this filter FIR4.

```
Variable name for impulse response :     FIR4
```

Impulse response can be plotted by pressing the [F2] key at this point (see Fig. 6-14). Analysis of the designed filter can be done as explained in previous sections. The log-magnitude of the frequency response is shown in Fig. 6-15.

Example 6-7:

Using the Parks-McClellan algorithm, design a 33-tap Hilbert transformer with optimum response in the normalized frequency range 0.05 to 0.45.

We will use the following parameter values:

```
Design method :    P          Filter type :    T

Filter order  :    33
```

For a Hilbert transformer, there is only one frequency band, and the program will not produce any prompts for the number of bands. Also, the first and the last edge frequencies are not required to be 0.0 and 0.5, as in the multiband filter case. Enter the edge frequencies, as follows:

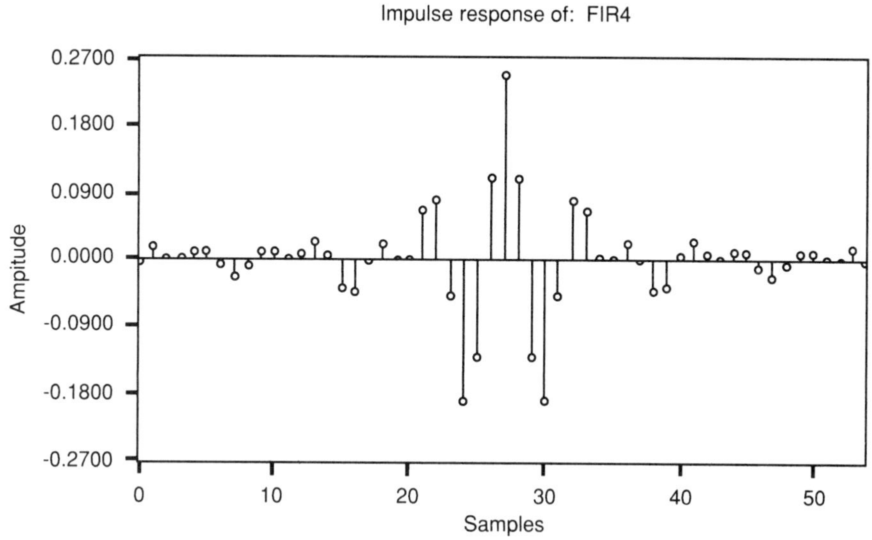

Figure 6-14. Impulse response of the filter designed in Ex. 6-6.

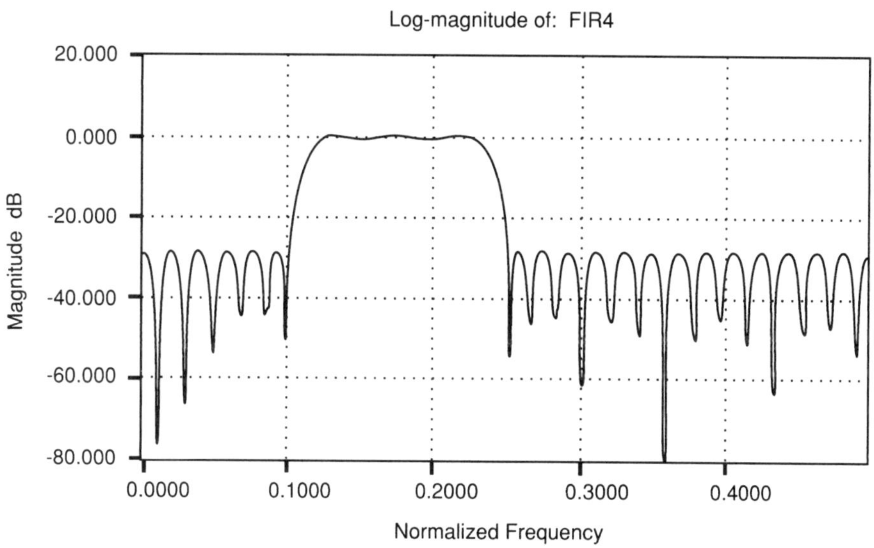

Figure 6-15. Log-magnitude of the filter designed in Ex. 6-6.

	Band 1
Lower Edge	0.05
Upper Edge	0.45

Use the name FIR5 for the impulse response. The magnitude of the frequency response for the designed Hilbert transformer is shown in Fig. 6-16.

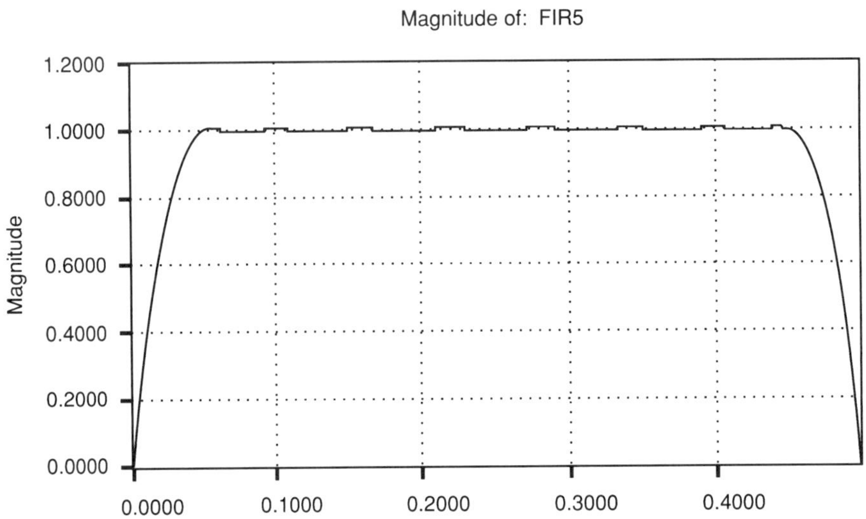

Figure 6-16. Magnitude response of the Hilbert transformer designed in Ex. 6-7.

6-4 LEAST SQUARES DESIGN OF FIR FILTERS

The third FIR filter design method implemented in PC-DSP is the least squares design method. This is basically a variant of the frequency sampling design method, extended to accept a transition band in the shape of a raised-cosine function. As an example, consider the following frequency response for a low-pass filter:

$$A(F) = \begin{cases} 1, & 0 < F < F_1 \\ 0.5 + 0.5\cos\left[\frac{(F - F_1)}{F_2 - F_1}1\right], & F_1 < F < F_2 \\ 0, & F_2 < F < 0.5 \\ A(1 - F), & 0.5 < F < 1.0 \end{cases}$$

To design an N-tap FIR filter, a sequence $H(k)$ is formed by sampling the desired response $A(F)$ at N equally spaced frequencies, i.e.,

$$H(k) = A\left(\frac{kF}{N}\right), \qquad k = 0, \ldots, N - 1$$

The N-point inverse DFT of $H(k)$ is computed and shifted to obtain the impulse response sought. Figure 6-17 illustrates the procedure for $N = 16$.

The least squares method can be used to design multiband filters with up to five disjoint frequency bands. The edge frequencies and the desired magnitude values are specified in exactly the same way as in the Parks-McClellan design.

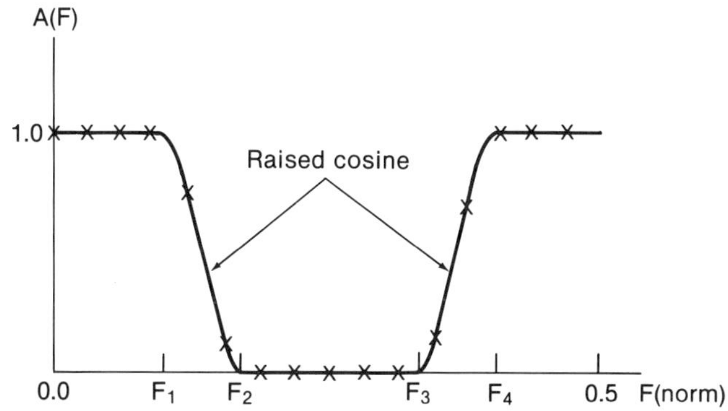

Figure 6-17. Least squares design with a transition region.

Example 6-8:

Using the least squares design method, design a 51-tap linear-phase low-pass FIR filter with band edges $F_1 = 0.2$ and $F_2 = 0.3$.

Select the *Design* function of the *FIR-filters* submenu, and make the following entries:

```
Design method :      L

Filter order  :      51              # of Bands :      2

                 Band 1         Band 2
Lower Edge       0.0                   0.3

Upper Edge       0.2                   0.5

Value            1                      0

Variable name for impulse response :      FIR6
```

The magnitude response of the designed filter is shown in Fig. 6-18. Note the raised-cosine shape of the transition region between 0.2 and 0.3.

The least squares design method can also be used to design traditional frequency sampling FIR filters without a transition band. The trick is to specify the band edge frequencies in such a way that no frequency samples fall into the transition band. The next example illustrates this point.

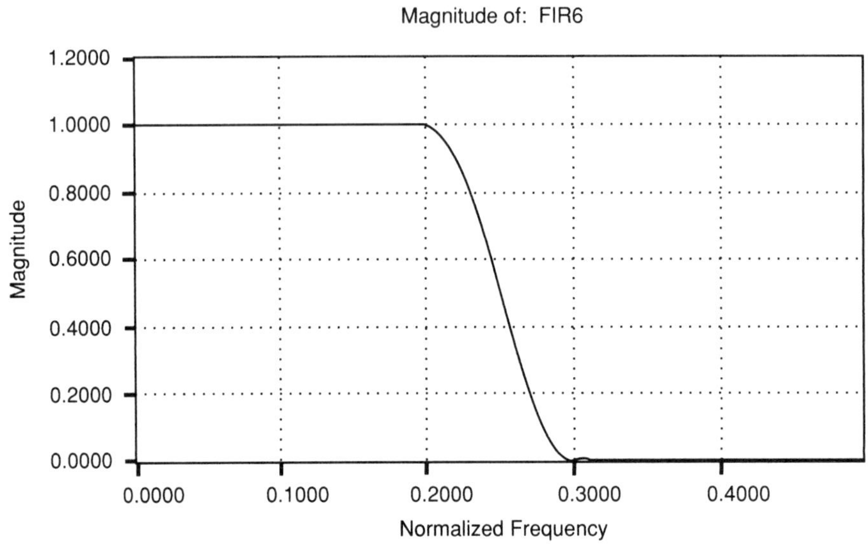

Figure 6-18. Magnitude response of the filter designed in Ex. 6-8.

Example 6-9:

Design a 21-tap frequency sampling FIR low-pass filter with a passband cutoff frequency $F_1 = 0.2$.

Refer to Fig. 6-17. For an N-tap filter, the frequency samples used in the design are spaced $1/N$ apart. With the proper choice of edge frequencies, we can ensure that all frequency samples are ones and zeros, with no samples taken in the raised-cosine transition band. We will design a two-band filter with the following edge frequencies:

	Band 1	Band 2
Lower Edge	0.0	0.21
Upper Edge	0.2	0.5
Value	1	0

The last frequency sample in the passband is at the normalized frequency $\frac{4}{21} = 0.1904$, and the first frequency sample in the stopband is at $\frac{5}{21} = 0.2380$. The choice of $F_1 = 0.2$ and $F_2 = 0.21$ ensures that no frequency samples are taken in the transition band. We will store the impulse response of the designed filter in the variable FIR7.

Example 6-10:

Redesign the FIR filter in Ex. 6-9. This time, use one transition sample with the value of 0.5 at the normalized frequency $\frac{5}{21}$. Compare the magnitude response with that of the filter in Ex. 6-9.

To get a value of 0.5 at the normalized frequency $\frac{5}{21}$, we have to ensure that this is the midpoint frequency of the raised-cosine transition band. We will also have to choose the edge frequencies of the transition band close enough so that only one frequency sample falls within. $F_1 = 4.5/21$ and $F_2 = 5.5/21$ satisfy both requirements.

	Band 1	Band 2
Lower Edge	0.0	5.5/21
Upper Edge	4.5/21	0.5
Value	1	0

We will call this filter FIR8. Figure 6-19 gives a comparison of the magnitude responses of FIR7 (solid line) and FIR8 (dashed line). It was generated by overlaying two magnitude responses in the *Graphics* component. It is apparent from the figure that the use of one transition sample greatly reduces the oscillatory behavior of the magnitude response.

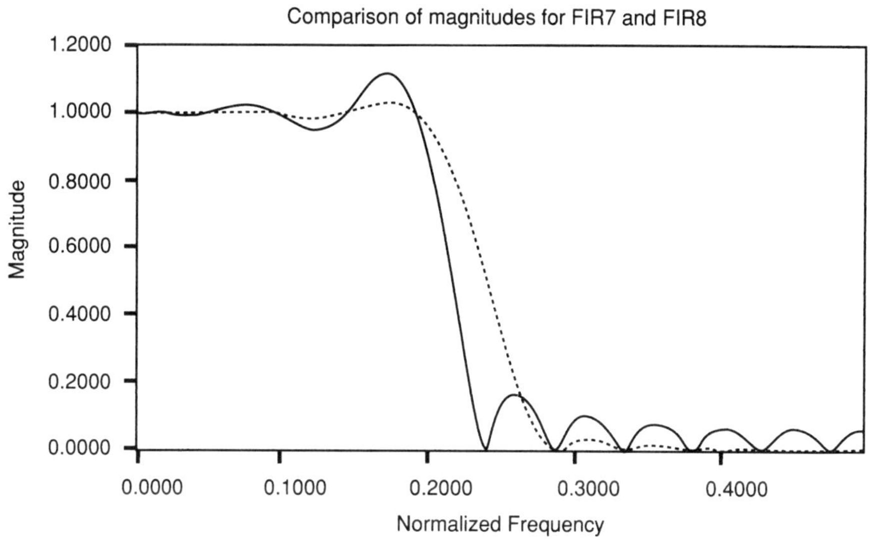

Figure 6-19. Comparison of results for Exs. 6-9 and 6-10.

6-5 EXERCISES

1. Compute and plot the dB magnitude of the frequency response for (a) rectangular, (b) triangular, (c) Hamming, and (d) Hanning windows. Use $M = 50$ points. For each window function, find the half-width of the main lobe and the decibel attenuation of the first side lobe.
2. Consider a three-tap averaging filter described by the difference equation

$$y(n) = \frac{1}{3}[x(n) + x(n-1) + x(n-2)]$$

a. Compute and plot the magnitude and the log-magnitude of the frequency response for this filter.

b. Process the noisy sawtooth waveform obtained in Ex. 3-11 through the three-tap averaging filter. Plot the output sequence. Comment on the result.

c. Repeat parts (a) and (b) for a five-tap averaging filter described by the difference equation

$$y(n) = \frac{1}{5}[x(n) + x(n-1) + x(n-2) + x(n-3) + x(n-4)]$$

3. Using the Fourier series design method with a rectangular window, design a 31-tap FIR filter with a normalized cutoff frequency $F_1 = 0.22$. Plot the impulse response and the magnitude response.

4. Repeat Exercise 3 with (a) a triangular window and (b) a Hamming window. How does the window affect the cutoff frequency? How does it affect the stopband response?

5. Using the Fourier series method with a rectangular window, design a 55-tap bandpass filter with normalized cutoff frequencies $F_1 = 0.18$ and $F_2 = 0.33$. Compute and plot the decibel magnitude.

6. Repeat Exercise 5 with (a) a triangular window and (b) a Hamming window. How does the choice of the window function affect the side-lobe attenuation? Can the edge frequencies be controlled precisely?

7. Design the FIR filter of Exercise 5 using the Parks-McClellan algorithm. Specify edge frequencies as follows:

Band 1	0.0 to 0.15
Band 2	0.18 to 0.33
Band 3	0.36 to 0.5

Use uniform weighting in each band. Plot the decibel response and compare with the filters designed in Exercises 5 and 6.

8. Using the Parks-McClellan algorithm, design a 25-tap FIR differentiator with unit slope. Choose the frequency band of interest between normalized frequencies 0.05 and 0.45. Plot the impulse response and the magnitude.

9. Generate 300 samples of the sinusoid

$$x(n) = 3\sin(0.2n), \qquad n = 0, \ldots, 299$$

and process through the FIR differentiator designed in Exercise 8. Compare the result with the theoretical derivative of $x(n)$. (*Hint*: Don't forget to take the 12-sample delay of the FIR filter into account.)

10. Repeat Exercise 9 using a periodic pulse train instead of a sinusoid. Generate 10 periods of the pulse train, with each pulse consisting of 15 ones and 10 zeros.

11. Using the least squares method, design a 21-tap notch filter to remove the normalized frequency $F = \frac{6}{21}$. Arrange the band edge frequencies such that only the frequency sample at $\frac{6}{21}$ is zero and all others are ones. Compute and plot the magnitude of the filter.

12. Generate 200 samples of the sequence

$$x(n) = \sin\left[2\pi\left(\frac{6}{21}\right)n\right]$$

and process through the filter designed in Exercise 11. Comment on the result.

7

Infinite Impulse Response Filters

This chapter concentrates on design, analysis, and implementation of infinite impulse response (IIR) digital filters. We also discuss methods for analyzing and implementing filters not designed by PC-DSP, such as filter designs carried out manually, taken from textbooks, or produced by other programs.

There are a number of methods for designing IIR filters. PC-DSP uses the popular bilinear transformation method. This is a three-step procedure. A typical design problem starts with a set of frequency domain specifications to be met by the digital filter. In the first step, the specifications of the desired digital filter are translated into the specifications of a corresponding analog prototype filter. In the second step the s-domain transfer function, $G(s)$, of this analog prototype is determined. Finally, in the third step $G(s)$ is transformed into a digital filter transfer function, $H(z)$, by means of the bilinear transformation defined as

$$s = \frac{2}{T}\frac{z-1}{z+1} \tag{7-1}$$

where T is the sampling interval.

7-1 ENTERING IIR FILTER SPECIFICATIONS

Before going through IIR filter design examples, we need to establish the terminology and notational conventions used in describing frequency domain behavior of IIR filters.

In general, an IIR filter is described by a rational z-domain transfer function $H(z)$. If this transfer function is evaluated on the unit circle of the z-plane, i.e., $z = e^{j\omega}$, we obtain

$$H(e^{j\omega}) = H(z)\big|_{z=(e^{j\omega})} = H(e^{j\omega})|e^{j\theta(\omega)} \tag{7-2}$$

where $|H(e^{j\omega})|$ and $\theta(\omega)$ are the magnitude and the phase response of the filter, respectively. In problems involving the design of IIR filters in the frequency domain, filter specifications are usually given in terms of magnitude only. This does not mean that the phase response is unimportant, but it is simply due to the fact that well-developed design techniques do not exist for designing IIR filters from phase specifications.

We will consider the specifications for a low-pass digital filter, depicted graphically in Fig. 7-1. The magnitude response of the filter is required to stay within unshaded areas. The range of frequencies from 0 to ω_1 radians is called the *passband* of the filter. In contrast, the range of frequencies from ω_2 to π radians is called the *stopband*. The parameters δ_1 and δ_2 are the *passband tolerance* and the *stopband tolerance*, respectively. Between the passband and the stopband there is a third frequency band called a *transition band* (also called a *don't-care band*) in the range from ω_1 to ω_2 radians.

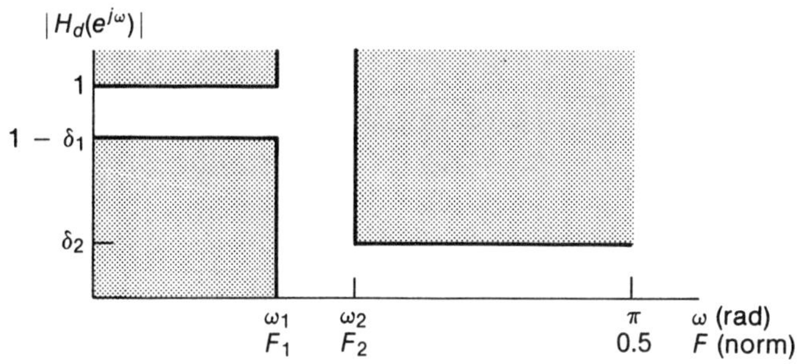

Figure 7-1. Magnitude specifications for a low-pass filter.

The magnitude plot can also be given in terms of the normalized frequency variable F, which is related to ω by the equation

$$\omega = 2\pi F \tag{7-3}$$

If the independent variable ω extends from 0 to π radians, then the corresponding normalized frequency variable F extends from 0.0 to 0.5. Another way to look at normalized frequencies is to treat them as fractions of the sampling frequency. For example, if a 300-hertz continuous sinusoid is sampled with a sampling rate of 1 kilohertz, then the normalized frequency of the resulting discrete sinusoid is 0.3. Note that normalized frequencies are dimensionless quantities, since they are obtained through division of actual frequencies in hertz by the sampling frequency, also in hertz.

In PC-DSP, filter specifications are entered using the normalized values of the edge frequencies involved. Thus, once the frequency axis is normalized, the solution of the design problem is independent of the actual value of the sampling frequency. This approach also makes it possible to have some uniformity in presenting the analysis results for the designed filter. If desired, however, the frequency axis may be denormalized after designing the filter to see the analysis plots in terms of actual frequencies.

The log-magnitude response of the filter (in decibels) is obtained through the equation

$$A_{dB}(\omega) = 20\log_{10}[|H(e^{j\omega})|] \tag{7-4}$$

The magnitude specifications shown in Fig. 7-1 can be easily converted to log-magnitude specifications by taking the logarithm of vertical axis values and scaling by a factor of 20 (see Fig. 7-2). The maximum allowed decibel passband tolerance is defined as

$$R_p = 20\log(1) - 20\log(1-\delta_1) = 20\log\left(\frac{1}{1-\delta_1}\right) \tag{7-5}$$

and the minimum required decibel stopband attenuation is

$$A_s = 20\log(1) - 20\log(\delta_2) = -20\log(\delta_2) \tag{7-6}$$

PC-DSP accepts both linear and logarithmic (dB) magnitude specifications. The examples of the next section will illustrate how magnitude specifications are entered into the program.

The definitions presented for the low-pass filter of Figs. 7-1 and 7-2 apply to other filter types as well. Specification diagrams for high-pass, band-pass, and band-reject filters are given in Fig. 7-3.

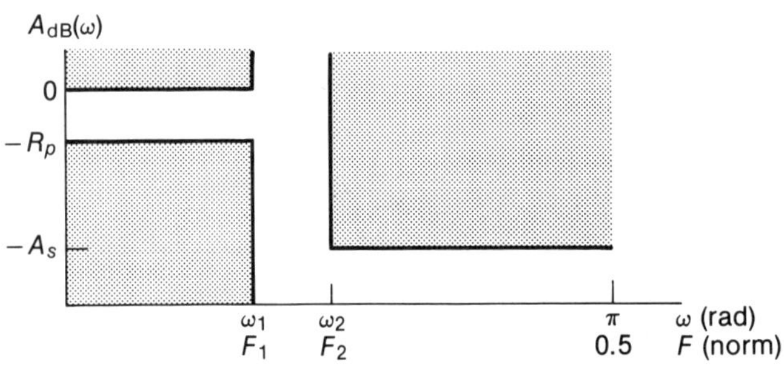

Figure 7-2. Log-magnitude specifications for a low-pass filter.

7-2 IIR FILTER DESIGN

In this section, we use two example IIR filter design problems to illustrate IIR filter design capabilities of PC-DSP.

Example 7-1

Design a fifth-order Butterworth low-pass filter to satisfy the following:

$$0.9 < |H(e^{j\omega})| < 1.0 \qquad \text{for } 0.0 < F < 0.2$$

Begin by selecting the *IIR-Filters* submenu and then the *Design* function. The first prompt to be answered is

```
Filter type   :    L
```

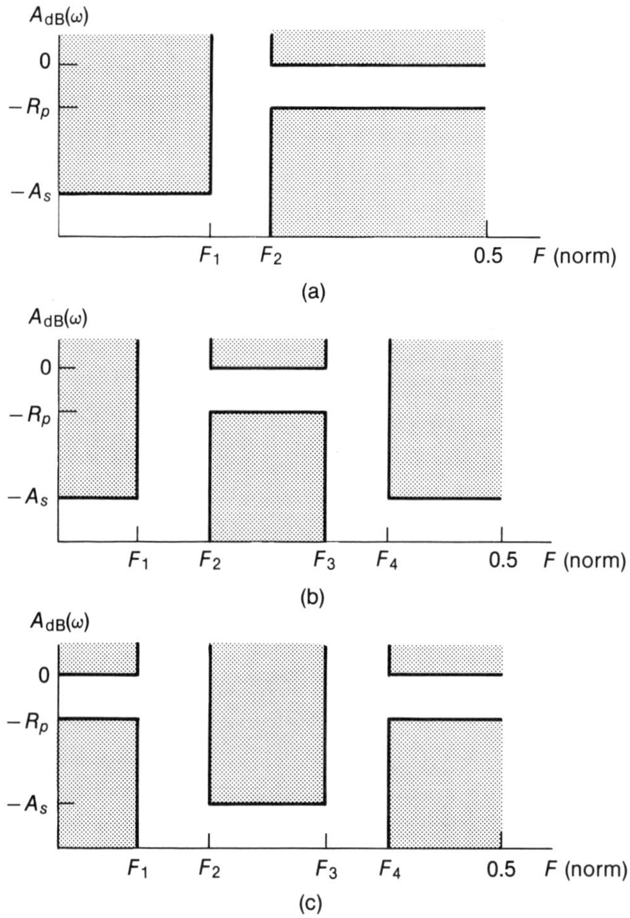

Figure 7-3. Specification diagrams for IIR filters: (a) highpass; (b) bandpass; (c) band-reject.

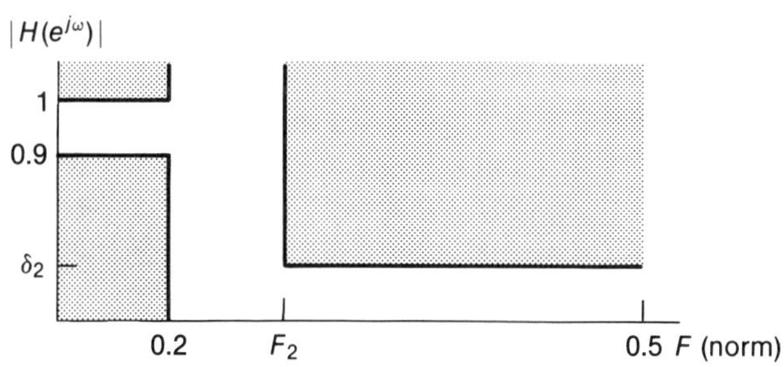

Figure 7-4. Specifications for the low-pass filter of Ex. 7-1.

Possible answers are listed in the adjacent help box,

```
L  =  Low-pass
H  =  High-pass
B  =  Band-pass
R  =  Band-reject
```

and the default type L (for low-pass) is suggested. Since we are designing a low-pass filter, press <ENTER> to accept the default. In the next step, we will specify the type of approximation to be used. The program prompt is

```
Approximation :    B
```

Available choices are again listed in the help box,

```
B = Butterworth
C = Chebyshev-I
I = Chebyshev-II
```

and the Butterworth approximation is suggested as the default. Again, press <ENTER> to accept it. The next entry is related to the way passband and stopband tolerances are specified.

```
Spec. type   :    1
```

The two possible choices are

```
1 = Design for a given filter order
2 = Determine filter order automatically
```

and the first one is suggested as the default choice. At this point, some background information is in order. Basically, in IIR filter design, there are three parameters that are dependent on each other, namely, the passband tolerance (δ_1 or R_p), the stopband tolerance (δ_2 or A_s), and the filter order N. Only two of these three parameters can be specified by the user, and the third parameter is determined by the program as a function of the first two.

If the first option in the help box is chosen, then the filter order can be specified by the user. When the design problem specifically calls for a fifth-order Butterworth filter, this is the proper choice. In this case, the next action of the program will be to ask for N. After entering N, the user can specify only one of the remaining two parameters. Which one is to be specified depends on the type of approximation used. (δ_1 or R_p for Butterworth and Chebyshev type-I filters, δ_2 or A_s for Chebyshev type-II filters.)

If the second option is chosen, then both tolerances are specified by the user. In this case, N is determined by the program as the smallest integer that satisfies the tolerance specifications.

Since the design problem specifies the filter order, answer the specification type prompt with a 1. After that, enter the filter order and the passband tolerance in response to the next two questions:

```
Filter order        :    5

Passband tolerance  :    0.1
```

Finally, enter the band edge frequencies. Recall from Section 7.1 that all edge frequencies are to be specified in terms of normalized values.

```
F1 :    0.2                 F2 :    0.3
```

Note that the stopband edge frequency F_2 is irrelevant to the design problem, since no stopband tolerance value is specified. Nevertheless, a value for F_2 must be entered on the prompt line, and any value that is greater than F_1 can be used. Thus, we have entered all filter specifications. The next entry is related to the printer report.

```
Printer report?  :  N    | Y = Yes |
                         | N = No  |
```

Answer this question with a Y if you want the program to generate a detailed report of the design procedure. Otherwise, accept the default answer N. The program should take a few seconds to design the filter and return with the following prompt

```
Variable name for filter data :
```

Now that the design procedure is complete, the program needs the name of the variable that will hold the description of the designed filter. This feature makes it possible to keep more than one filter design in memory. A number of filter designs can be stored with distinct variable names. They can be recalled for analysis and comparison. Enter the variable name

```
IIR1
```

for the filter we have just designed. The program will respond with the message

```
Variable = IIR1     , Type = C , Size =   13
```

The filter description is stored as a complex sequence with 13 samples. This completes the design procedure. We will look at the specific contents of the sequence IIR1 in Section 7.4. As you can see, there is no automatic analysis of the designed filter. Instead, the analysis is performed at will by selecting the *Analyze* function of the *IIR-filters* submenu. This is done so that filter designs obtained from textbooks or other design programs can be analyzed in exactly the same way as filters designed using PC-DSP. To analyze a filter with a known transfer function, we can create the filter description variable by using the *External* function of the *IIR-filters* submenu and then proceed as if that filter was designed by PC-DSP.

Example 7-2

Design a band-pass Chebyshev filter to satisfy the following specifications:

$$R_p = 1\text{dB} \qquad \text{for } 300\text{Hz} < f < 500\text{Hz}$$

$$A_s = 20\text{dB} \qquad \text{for } f < 200\text{Hz}, \text{ or } f > 650\text{Hz}$$

The sampling frequency is $f_s = 1600$ hertz. Meet the specifications with the lowest filter order possible.

This design problem differs from the previous one in several ways. Edge frequencies are specified in terms of actual hertz frequencies rather than normalized frequencies. Also, passband and stopband tolerances are given in decibels. Select the *Design* function from the *IIR-Filters* submenu, and answer prompts as follows:

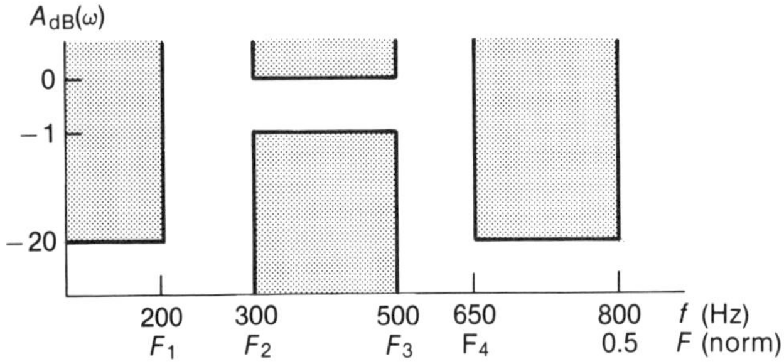

Figure 7-5 Specifications for the bandpass filter of Ex.7-2

```
Filter type    :    B          Approximation :    C

Spec. type     :    2

Passband tolerance :    1d

Stopband tolerance :    20d

F1 :    200/1600          F2 :    300/1600

F3 :    500/1600          F4 :    650/1600

Printer report :    Y        ( Answer N if no printer )

Variable name for filter data :    IIR2
```

Note that the decibel tolerance values are followed by the letter *d*. This is how PC-DSP distinguishes between linear and logarithmic values. Also, it is not necessary to convert actual frequencies to normalized frequencies manually. Instead, we can just enter arithmetic expressions for normalized frequencies and let PC-DSP evaluate them.

The printer report generated by PC-DSP is instructive about the design process, as shown in Fig. 7-6. The first part of the report is an echo of the user's specifications. PC-DSP then computes the prewarped edge frequencies $\omega_1, \omega_2, \omega_3$, and ω_4 of the corresponding analog bandpass prototype filter from the normalized edge frequencies of the digital filter.

Since an analog band-pass filter is designed from an analog low-pass filter through frequency transformations, the next step is to compute the edge frequencies ω_p and ω_s of a proper analog lowpass prototype.

```
===== PC-DSP    IIR Filter Design Report =====

   Date : 03-19-1989
   Time : 12:49:57

   Filter type     : Bandpass
   Approximation   : Chebyshev-I

   Passband tolerance (dB)    : 1
   Stopband attenuation (dB)  : 20

   Nomalized edge frequencies:
   F1 = .125
   F2 = .1875
   F3 = .3125
   F4 = .40625

   Pre-warped edge frequencies for the
   analog prototype (in rad/sec):
   W1 = .8284272
   W2 = 1.336357
   W3 = 2.993212
   W4 = 6.59312

   Edge frequencies for the analog lowpass
   prototype (in rad/sec):
    WP = 1.656855
    WS = 4.000002

   Min. filter order = 6

   Epsilon = .5088471

   S-domain poles of the analog lowpass prototype:

   -4.093845E-01 + j  1.600519E+00
   -8.187690E-01 + j  0.000000E+00
   -4.093845E-01 + j -1.600519E+00

   S-domain poles of the analog prototype:

   -2.810318E-01 + j  2.946033E+00
   -4.093848E-01 + j  1.957653E+00
   -1.283534E-01 + j  1.345514E+00
   -2.810312E-01 + j -2.946033E+00
   -4.093842E-01 + j -1.957653E+00
   -1.283527E-01 + j -1.345514E+00

   S-domain zeros of the analog prototype:

    0.000000E+00 + j  0.000000E+00
    0.000000E+00 + j  0.000000E+00
    0.000000E+00 + j  0.000000E+00

   Z-domain poles of the digital filter:

   -3.427472E-01 + j  8.488653E-01
   -2.617841E-07 + j  8.125115E-01
    3.427469E-01 + j  8.488652E-01
   -3.427473E-01 + j -8.488654E-01
   -1.616692E-07 + j -8.125117E-01
    3.427471E-01 + j -8.488656E-01

   Z-domain in zeros of the digital filter:

    1.000000E+00 + j  0.000000E+00
    1.000000E+00 + j  0.000000E+00
    1.000000E+00 + j  0.000000E+00
   -1.000000E+00 + j  0.000000E+00
   -1.000000E+00 + j  0.000000E+00
   -1.000000E+00 + j  0.000000E+00

   Constant gain factor = 2.107472E-02
```

Figure 7-6. Printer report for the design problem in Ex. 7-2.

The minimum required filter order is computed and printed. The parameter ε for the analog low-pass Chebyshev filter is also shown. The poles of the analog low-pass filter are computed and listed. Using the proper transformation, poles and zeros of the analog band-pass prototype filter are computed. Finally, poles and zeros of the digital filter are determined using a bilinear transformation.

Details of the pole-zero calculations and the frequency transformations involved are beyond the scope of this workbook and can be found in most textbooks that treat analog and digital filter design.

7-3 ANALYSIS OF DESIGNED FILTERS

We will now look at the IIR analysis options of PC-DSP. The analysis of the designed filter is done by selecting the *Analyze* function of the *IIR-filters* submenu. Program prompts are self-explanatory. We present two examples.

Example 7-3

Tabulate the poles and the zeros of the filter designed in Ex. 7-2. Obtain a pole-zero plot. Also generate coefficients for a cascade implementation of this filter with second-order blocks.

Select the *Analyze* function from the *IIR-filters* submenu. The program first asks for the name of the filter to be analyzed, and the filter IIR2 that we have just designed in Ex. 7-2 is suggested as the default answer.

```
Filter description variable :     IIR2
```

PC-DSP remembers the name of the IIR filter variable that was most recently accessed and uses it as the default. Just press the <ENTER> key to accept this default. Available analysis options are listed in a help box.

```
I = Impulse response
Z = Poles and Zeros
C = Cascade sections
M = Magnitude plot
L = Log-magnitude plot
P = Phase plot
G = Group delay plot
```

Since we wish to tabulate the poles and the zeros of the designed filter, we will choose Z. The program generates the list shown in Fig. 7-7.

If you have a printer connected, you can also obtain a hard copy of the poles and the zeros of the filter by pressing the [F3] function key. A pole-zero plot can be obtained by pressing the [F2] function key (see Fig. 7-8).

Now press the <ESC> key return to the *IIR-filters* submenu and select *Analyze* one more time. When prompted for the analysis type, enter a C for cascade section coefficients. PC-DSP now lists the coefficients of second order cascade sections, each in the form

$$H_k(z) = \frac{a_2 z^2 + a_1 z + a_0}{b_2 z^2 + b_1 z + b_0}$$

```
                                                          PC-DSP v1.1
  System  Data  Filter  sPectrum  Graphics  Macro  Quit

  FILTER / IIR-FILTERS / ANALYZE

  Filter description variable :  IIR2          ---> Type = C , Size =   15

  Analysis type :  Z

  Filter order  :  6                Gain factor :  2.10747E-02

             Zeros                                    Poles

   1.00000E+00 + j  0.00000E+00             3.42747E-01 + j -8.48865E-01
   1.00000E+00 + j  0.00000E+00            -2.61784E-07 + j  8.12512E-01
   1.00000E+00 + j  0.00000E+00             3.42747E-01 + j  8.48865E-01
  -1.00000E+00 + j  0.00000E+00            -3.42747E-01 + j -8.48865E-01
  -1.00000E+00 + j  0.00000E+00            -1.61669E-07 + j -8.12512E-01
  -1.00000E+00 + j  0.00000E+00            -3.42747E-01 + j  8.48865E-01

  F1: Help     F2: Plot     F3: Print     <ESC>: Menu
```

Figure 7-7. On-screen tabulation of poles and zeros.

Pole-zero plot for: IIR2

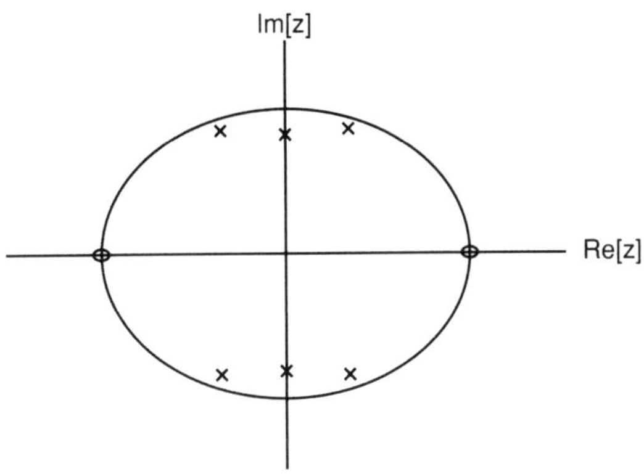

Figure 7-8. Pole-zero plot for the band-pass filter of Ex. 7-2.

```
                                                          PC-DSP v1.1
  System  Data  Filter  sPectrum  Graphics  Macro  Quit

  FILTER / IIR-FILTERS / ANALYZE

  Filter description variable :  IIR2          ---> Type = C , Size =   15

  Analysis type :  C

  Filter order  :  6                Gain factor :  2.10747E-02

      A2          A1          A0          B2          B1          B0

    1.0000     -2.0000      1.0000      1.0000      0.0000      0.6602
    1.0000      0.0000     -1.0000      1.0000     -0.6855      0.8380
    1.0000      2.0000      1.0000      1.0000      0.6855      0.8380

  F1: Help                  F3: Print     <ESC>: Menu
```

Figure 7-9. On-screen tabulation of the coefficients of second-order cascade blocks.

The transfer function of the filter IIR2 consists of three second-order blocks and a constant-gain factor, i.e.,

$$H(z) = KH_1(z)H_2(z)H_3(z)$$

where $K = 0.0210747$ and

$$H_1(z) = \frac{z^2 - 2z + 1}{z^2 + 0.6855z + 0.8380}$$

$$H_2(z) = \frac{z^2 - 1}{z^2 + 0.6602}$$

$$H_3(z) = \frac{z^2 + 2z + 1}{z^2 - 0.6855z + 0.8380}$$

Example 7.4

Compute and plot magnitude, phase, and group delay for the filter stored in IIR2.

Select *Analyze* again, and enter M in response to the analysis-type prompt. The program will compute the filter magnitude and, when finished, will prompt you for the name of the variable in which the magnitude is to be stored.

```
Variable name for analysis results :     TEMPR
```

The default name, TEMPR, is suggested. Press the <ENTER> key to accept the default name. The program will respond with

```
Variable = TEMPR    , Type = R , Size = 256
Variable = NFREQ    , Type = R , Size = 256
```

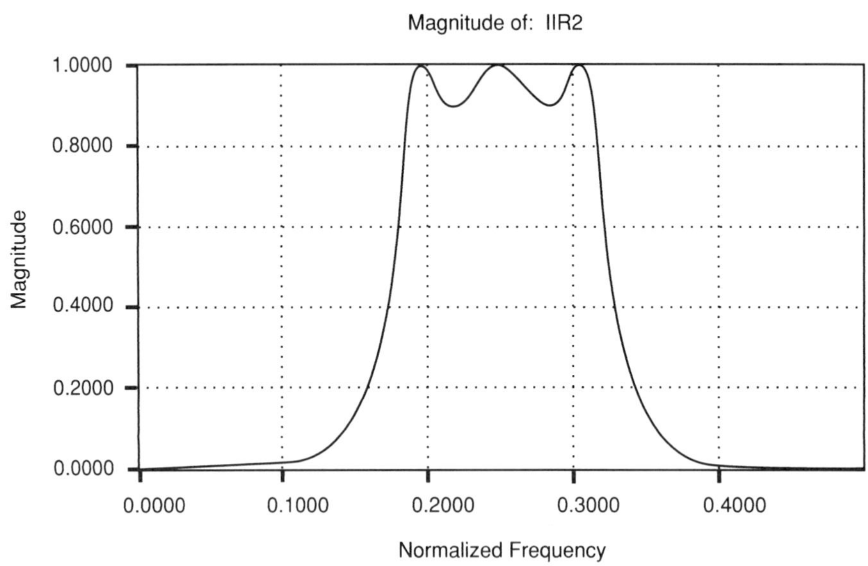

Figure 7-10. Magnitude response of the Chebyshev band-pass filter designed in Ex. 7-2.

Two variables are generated. TEMPR holds 256 points of the filter magnitude for normalized frequencies from 0 to 0.5. The second variable, NFREQ, holds the normalized frequencies that correspond to the magnitude samples in TEMPR. It is used only for plotting the magnitude as a function of frequency and may be deleted after looking at the plot. Now press the [F2] key to obtain the magnitude plot shown in Fig. 7-10.

The magnitude plot can be customized, if desired, by using the variables TEMPR and NFREQ in the *Graphics* menu. Figure 7-11 shows a blow up of the denormalized frequency range from 160 to 320 hertz. This plot was generated by first multiplying NFREQ by 1600 and then adjusting the parameters X-min and X-max in the *Graphics* menu. The density of the grid was changed by adjusting the data label increments.

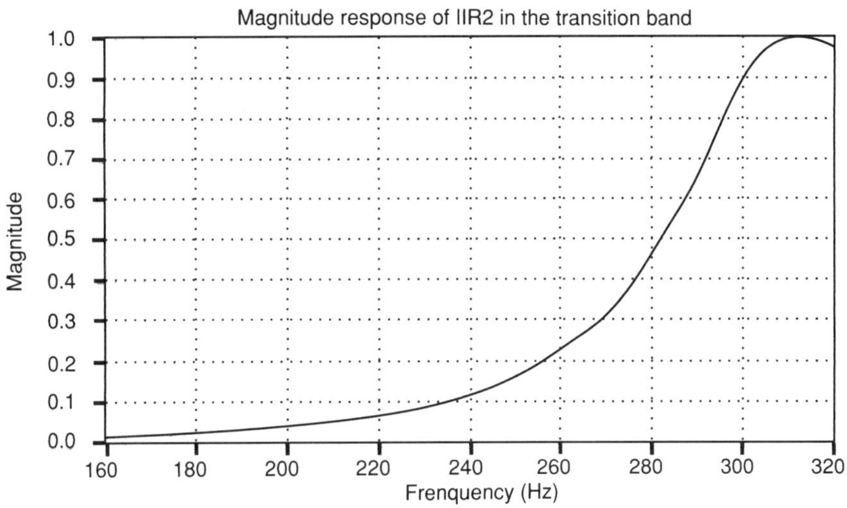

Figure 7-11. Magnitude response in the transition band.

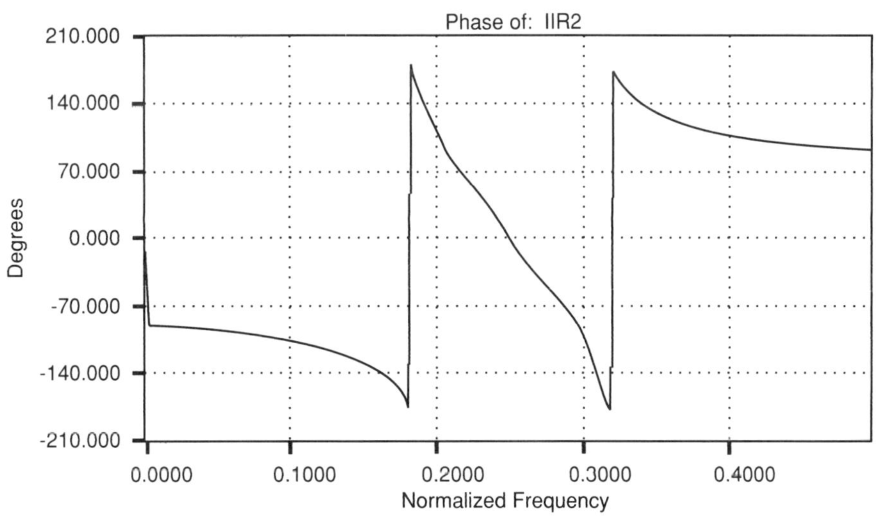

Figure 7-12. Phase response of the Chebyshev band-pass filter designed in Ex. 7-2.

Phase and group delay plots can be obtained in a similar way (see Figs. 7-12 and 7-13). Note that the phase is in degrees and is displayed in the range from -180 to 180. The group delay is in terms of sampling intervals. For this filter, the maximum group delay is a little over 12 sampling intervals, or $12/f_s$ seconds. The vertical axis can be changed to seconds by multiplying the group delay variable by the sampling interval T $(= 1/f_s)$.

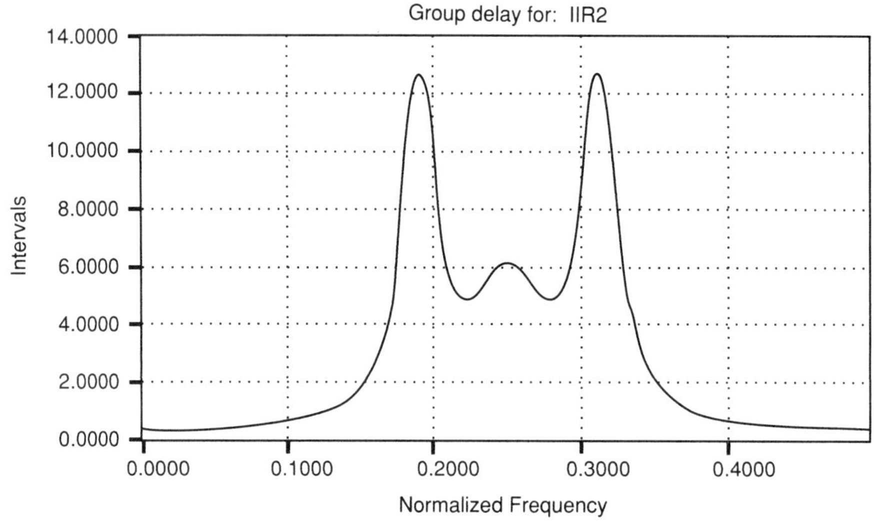

Figure 7-13. Group delay for the Chebyshev band-pass filter designed in Ex. 7-2.

7-4 ANALYZING EXTERNALLY DESIGNED FILTERS

One of the important features of PC-DSP is the capability to analyze and implement filters that are external to the program, such as filter designs taken from textbooks or pencil-and-paper designs of the user. This capability is made possible by storing the filter description in a named sequence that can be created, accessed, or modified just like any other PC-DSP variable.

To understand how the description of a filter is stored in a named sequence, we will use the variable IIR2 of the previous sections as an example and tabulate it on the screen using the *Print* function of the *System* menu (see Fig. 7-14). Note that the real part of element 0 holds the constant gain factor, 0.0218645. The next two elements of IIR2 are integers indicating the number of zeros (6) and the number of poles (6), respectively. After that, z-domain zeros of the filter transfer function are listed, followed by z-domain poles.

To analyze an external filter characterized by its zeros and poles, the filter description variable can be created manually using the *Read* function of the *System* menu.

Example 7-5

The following system function is borrowed from *Discrete Time Signal Processing* by A. V. Oppenheim and R. W. Schafer, ©1989 pp. 228–229. Adapted by permission of Prentice Hall, Inc., Englewood Cliffs, New Jersey.

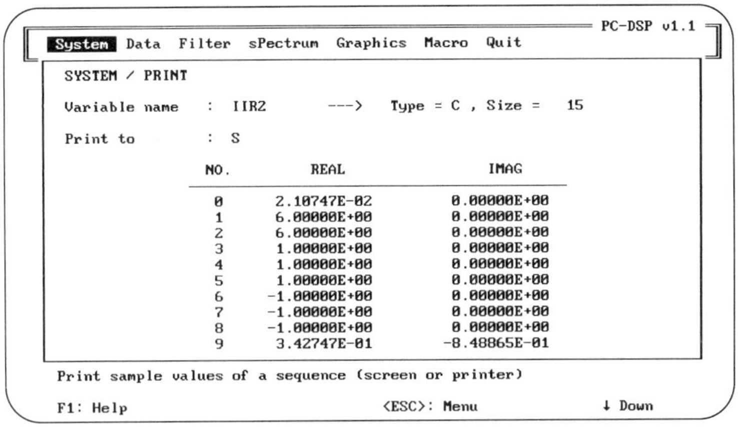

Figure 7-14. On-screen tabulation of the sequence IIR2.

Consider a linear system described by the z-domain transfer function

$$H(z) = \frac{0.05634(1 + z^{-1})(1 - 1.0166z^{-1} + z^{-2})}{(1 - 0.683z^{-1})(1 - 1.4461z^{-1} + 0.7957z^{-2})}$$

The zeros of the transfer function are at

Radius	Angle (rad)	Real part	Imaginary part
1.0	π	-1.0	0.0
1.0	∓ 1.0376	0.5083	∓ 0.8612

The poles are at

Radius	Angle (rad)	Real part	Imaginary part
0.683	0.0	0.683	0.0
0.892	∓ 0.6257	0.723	∓ 0.5224

Plot log-magnitude, phase, and group delay for this system.

Using the *Read* function of the *System* menu, create the filter description variable IIR3 with the following samples:

Sample Number	Value
0	0.05634
1	3
2	3
3	-1
4	$0.5083 + j0.8612$
5	$0.5083 - j0.8612$
6	0.683
7	$0.723 + j0.5224$
8	$0.723 - j0.5224$

Once the filter description variable is created, desired plots can be obtained by using the *Analyze* function of the *IIR-filters* submenu, as explained in the previous section (see Figs. 7-15 through 7-17).

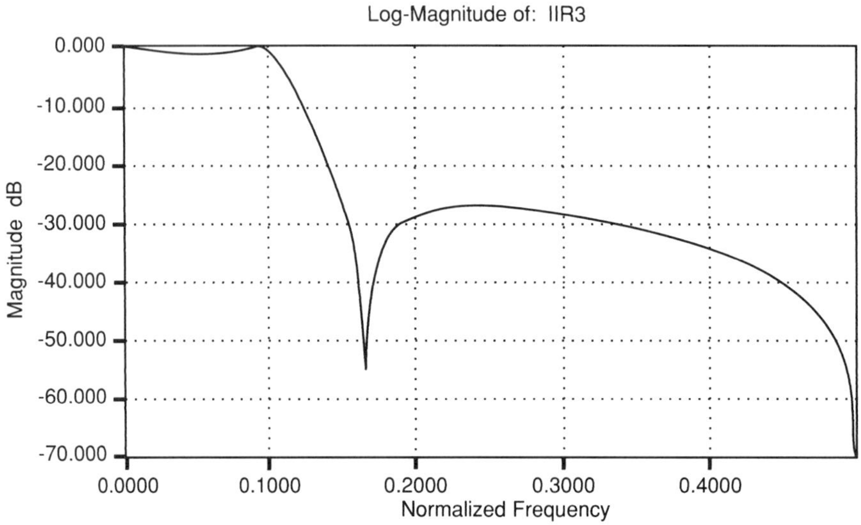

Figure 7-15. Log-magnitude for the system of Ex. 7-5.

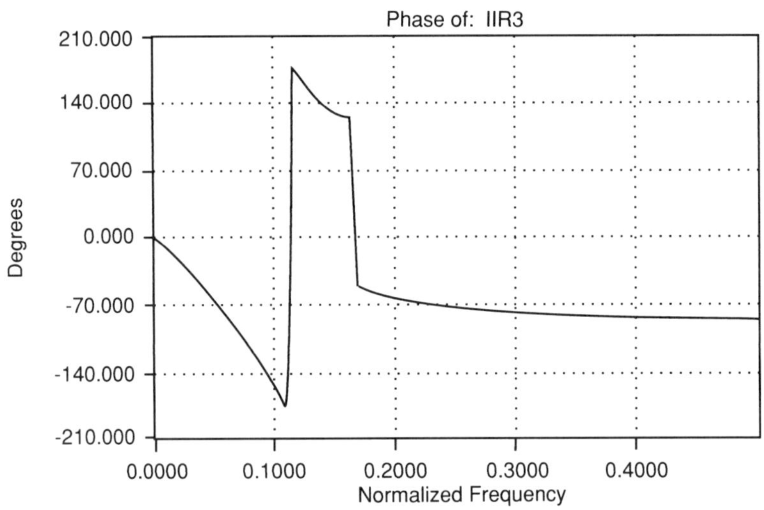

Figure 7-16. Phase response for the system of Ex. 7-5.

In the preceeding example, we have shown that for a filter described by its z-domain poles and zeros, the filter description variable can be created in a very straightforward manner. If the system is described by a rational transfer function instead of poles and zeros, the *External* function of the *IIR-filters* submenu can be used to create the filter description variable from the knowledge of numerator and denominator coefficients.

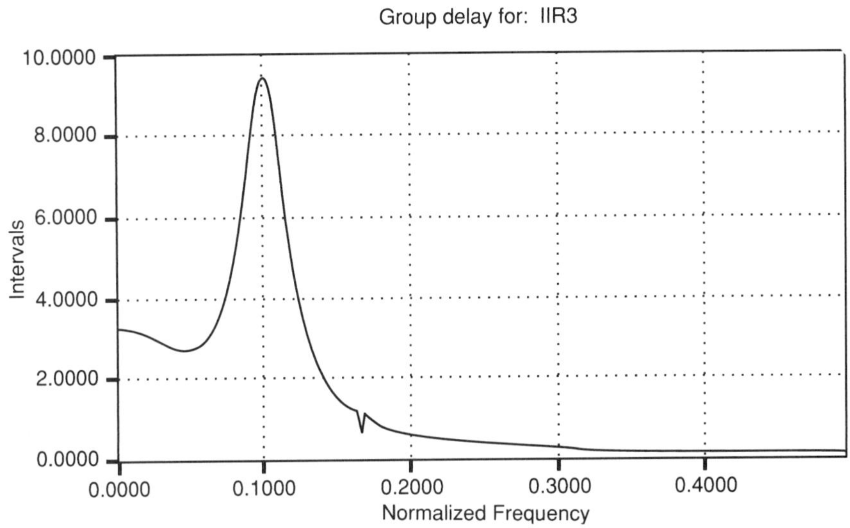

Figure 7-17. Group delay for the system of Ex. 7-5.

Example 7-6

Create the filter description variable for the IIR filter described by the z-domain transfer function

$$H(z) = \frac{z^2 + z}{z^5 - 1.45z^4 + 0.9z^3 + 0.39z^2 - 0.74z + 0.3}$$

First, using the *Read* function of the *System* menu, generate the following two variables:

```
HNUM = { 1   1   0 }
HDEN = { 1  -1.45   0.9   0.39   -0.74   0.3 }
```

HNUM and HDEN contain numerator and denominator coefficients of $H(z)$ ordered from the highest power of z to the lowest. Now select *External* and make the following entries:

```
Numerator   :    HNUM

Denominator :    HDEN

Variable name for filter data :    IIR4
```

IIR4 now contains a description of the filter in the PC-DSP format and can be used for analysis.

A practical tip: If you ever need a polynomial root finding routine, the *External* function can be used to factor a general polynomial. Just create a variable that contains the polynomial coefficients from the highest to the lowest power and use it as either the numerator or the denominator of a hypothetical system to obtain the filter description variable. Then select *Analyze* and tabulate zeros and poles.

7-5 IIR PROCESSING OF DISCRETE DATA

The *Process* function of the *IIR-filters* submenu is used for simulating an IIR filter. Any real or complex sequence that is either generated within PC-DSP or loaded from a data file can be used as the input signal, and the output sequence of the filter can be obtained. Internally, PC-DSP uses a simulation diagram that consists of second-order cascade blocks, with each block implemented in direct-form-II. The filter is assumed to be initially relaxed, i.e., all initial conditions are zero. We give an example.

Example 7-7

The filter IIR3 of Ex. 7-5 is driven by the sequence XN that is in the data file EX7-7.DSP. (This data file is in the Utilities and Examples Disk.) Compute and plot the output of the filter.

The input sequence loaded from the data file is a 140-sample segment from a periodic pulse train, as shown in Fig. 7-18. Now select the *Process* function from the *IIR-filters* submenu. A data entry window will appear at the bottom half of the screen. Make the following entries:

```
Variable name for filter data :      IIR3
Input  :     XN
Output :     YN
```

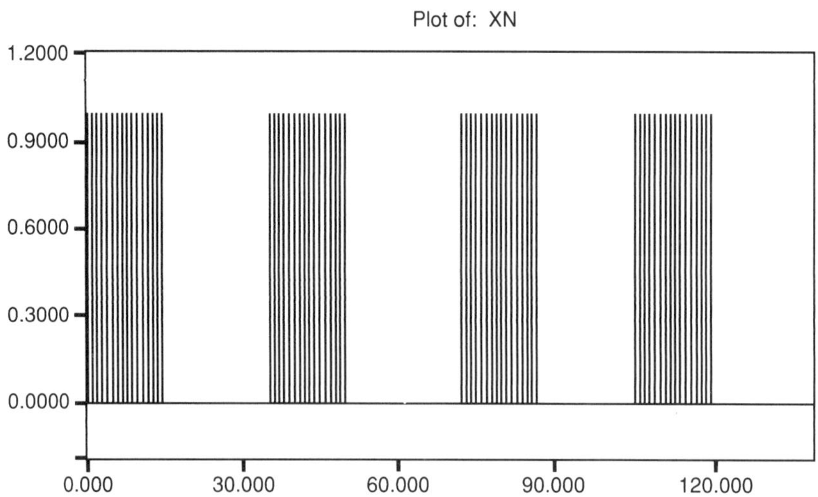

Figure 7-18. The input sequence for Ex. 7-7.

A quick-plot of the resulting output sequence can be obtained by pressing the [F2] function key and is shown in Fig. 7-19. The number of computed output samples is equal to the number of input samples. Note the transient in the first few samples of the response. This is due to zero initial conditions. Also note the smearing effect of the low-pass filter on the pulses.

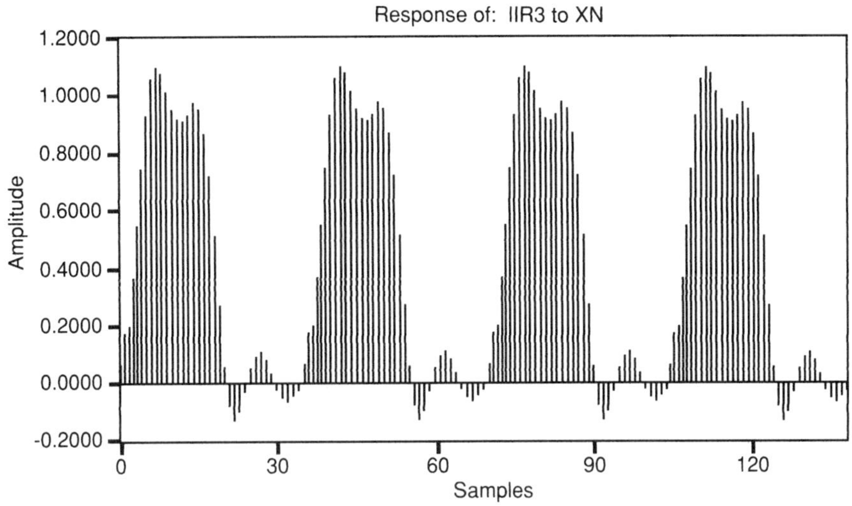

Figure 7-19 The output sequence for Ex. 7-7.

7-6 EXERCISES

1 Design a Butterworth low-pass digital filter to satisfy the following requirements:

$$0.99 < |H(e^{j2\pi F})| < 1 \quad \text{for} \quad 0 < F < 0.22$$
$$0 < |H(e^{j2\pi F})| < 0.01 \quad \text{for } 0.25 < F < 0.5$$

Design the lowest-order filter possible. What is the filter order? Plot the magnitude, phase, and group delay characteristics for the designed filter.

2 Tabulate the poles and the zeros of the z-domain transfer function for the filter designed in Exercise 1. Using the poles and the zeros, write the z-domain transfer function.

3 Tabulate the coefficients of second-order cascade sections for the filter designed in Exercise 1. Using these coefficients, draw a block diagram realization of the filter.

4 Repeat Exercises 1–3 for a Chebyshev type-I filter.

5 Design a tenth-order Chebyshev type-II band-reject filter to remove the normalized frequency $F = 0.22$. Choose the stopband edge frequencies such that the stopband is centered around 0.22 and its width is 0.04. Choose the widths of both transition bands to be equal to 0.03. Also, ensure at least 20-dB attenuation in the stopband. Use $R_p = 1$ dB.

6 Generate 300 samples of the sequence

$$x(n) = \sin[2\pi(0.22)n], \qquad n = 0, \ldots, 299$$

and process through the band-reject filter designed in Exercise 5. Comment on the output sequence.

7 A linear time-invariant system is described by the z-domain transfer function

$$H(z) = \frac{z}{z - 0.9}$$

a. Using the *External* function, generate the filter description variable for this system.

b. Compute and plot the magnitude and the phase of the frequency response. Note the magnitude and the phase at the normalized frequency 0.05.

c. Generate 500 samples of the sinusoid

$$x(n) = \sin[2\pi(0.05)n], \qquad n = 0, \ldots, 499$$

and process through the filter. Compare the steady-state portion of the response to the original input sequence. How are the amplitude and the phase of the sinusoid affected by the filter?

d. Repeat part (c) with the sinusoid

$$x(n) = \sin[2\pi(0.1)n], \qquad n = 0, \ldots, 499$$

8 Consider two continuous-time variables $x_a(t)$ and $y_a(t)$ that have an integral relationship

$$y_a(t) = \int_0^t x_a(t)dt$$

The integral can be approximated using the trapezoidal rule, as follows:

$$y_a(t) = y_a(t_0) + \frac{t - t_0}{2}[x_a(t) + x_a(t_0)]$$

A trapezoidal integrator can be represented by the difference equation

$$y(n) = y(n-1) + \frac{T}{2}[x(n) + x(n-1)]$$

where $y(n)$ and $x(n)$ are sampled versions of $y_a(t)$ and $x_a(t)$, respectively, i.e.,

$$y(n) = y_a(nT), \qquad x(n) = x_a(nT)$$

a. Determine the z-domain transfer function, $H(z)$, of the trapezoidal integrator.

b. Using the *External* function, create a filter description variable for the trapezoidal integrator. Use $T = 0.1$ s.

c. The signal $x_a(t)$ is given as

$$x_a(t) = (0.9)^t \sin(2t)$$

The integral is to be approximated using the trapezoidal rule. To do this, the signal $x_a(t)$ is sampled with $T = 0.1$ s intervals, and the resulting sequence $x(n)$ is processed through the trapezoidal integrator. Compute 100 samples of the resulting approximation to the integral.

d. Compute $y_a(t)$ analytically and compare it with the output of the trapezoidal integrator obtained in part (c). Compute and plot the error between the true and the approximate values of the integral.

e. Repeat parts (a) through (d) for $T = 0.05$ s.

9 Consider a linear time-invariant system with the system function

$$H(z) = \frac{1}{1 - z^{-N}}$$

In Exercise 4-10, it was mentioned that this system could be used for making a sequence periodic with N.

a. Using $N = 10$, create a filter description variable for this system. Then use the *Analyze* function to generate 100 samples of the impulse response.

b. Compute and plot the magnitude and the phase of the frequency response for this system.

c. Generate a 10-sample sequence

$$x(n) = 9 - n, \qquad n = 0, \ldots, 9$$

Pad $x(n)$ with 90 zeros. Process it through the system. Compute and plot 100 samples of the output sequence.

8

Spectrum Analysis

PC-DSP implements five popular spectrum analysis algorithms. The first two of these algorithms, Welch's method of averaging modified periodograms and the Blackman-Tukey method, are classical FFT-based algorithms. The other three are time-series modeling techniques using autoregressive (AR), moving-average (MA), and autoregressive-moving-average (ARMA) models. In this chapter, mathematical details and usage of each method are explained. It should be kept in mind, however, that this text does not attempt to give a thorough treatment of the theory of spectrum analysis. The user is advised to learn the theory behind each spectrum analysis algorithm from textbooks prior to going through this chapter.

In the spectrum analysis examples of this chapter, we use a real sequence XREAL and a complex sequence XCOMP. These sequences are in data files EX8-1.DSP and EX8-2.DSP, respectively. Each sequence has 100 samples. The sequence XCOMP was generated as

$$x_{\text{comp}}(n) = \exp[j2\pi(0.2)n] + \exp[j2\pi(0.208)n] + w(n)$$

where $w(n)$ is a complex Gaussian white noise process in the form

$$w(n) = w_r(n) + jw_i(n)$$

Both $w_r(n)$ and $w_i(n)$ were generated using the *Normal* function with the mean equal to zero and the variance equal to 0.2. The real sequence $x_{\text{real}}(n)$ was obtained by simply taking the real part of $x_{\text{comp}}(n)$.

8-1 WELCH'S METHOD OF AVERAGING MODIFIED PERIODOGRAMS

The periodogram estimator of Power Spectral Density (PSD) is defined as

$$P_{\text{per}}(F) = \frac{1}{N}\left|\sum_{n=1}^{N-1} x(n)\exp(-j2\pi Fn)\right|^2 \tag{8-1}$$

where $x(n)$ represents the available N-point data record from the process for which the PSD is to be estimated. A comparison of the magnitude of the squared term in Eq. (8-1) and the FFT definition in Eq. (4-1) suggests that it is possible to use FFT to compute the periodogram in a fast way. This is indeed how it is computed in PC-DSP.

It can be shown that the periodogram is not a consistent spectral estimator. The variance of the estimate does not tend to zero as the record length approaches infinity, i.e., $N \to \infty$. One method has been proposed by Welch [1967] for improving the statistical properties of the periodogram. The Welch method is based on breaking up the N-point data record $x(n)$ into M-point segments $x_k(n)$ that overlap with each other by L samples.

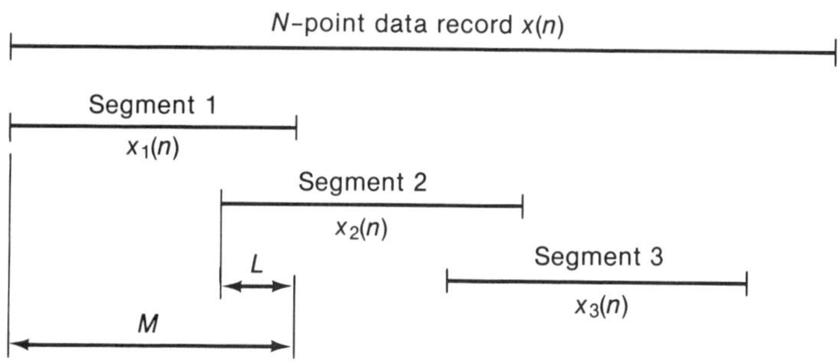

Figure 8-1. Segmentation of the data record for the Welch method.

After the data record is broken into overlapping M-point segments, a data window is applied to each segment. Then, a periodogram is computed for each windowed segment using the definition given in Eq. (8-1). Finally, these periodograms are averaged, and the result is scaled to obtain the Welch estimate. To distinguish between the periodogram and its variant due to Welch, we will use the names *periodogram* and *Welch periodogram*, respectively.

The *Periodogram* function of the *Spectrum* component can be used for obtaining both the standard periodogram and the Welch periodogram. Three examples are given to illustrate this function.

Example 8-1:

Compute and plot the decibel periodogram for the real data sequence XREAL.

Begin by loading the sequence XREAL from the data file EX8-1.DSP. Select the *Spectrum* component and then select *Periodogram*.

First, the variable that holds the input data record is specified by answering the prompt

```
Variable name for input data :    XREAL
```

Since we want to compute the standard periodogram, we will choose the segment length to be equal to the length of the sequence.

```
Segment length  :    100
```

This has the effect of using the entire input sequence as one segment. The next parameter to be entered is the number of overlapping samples between segments.

```
Segment overlap :    0
```

The default value is 0. Since we are using only one segment (which is the entire sequence), this last parameter has no effect in the calculations to be performed. Just accept the default value by pressing the <ENTER> key. Next, PC-DSP will prompt you for the type of window function to be used on each segment.

```
Window type :    R
```

Possible window choices are listed in a help box.

```
R  =  Rectangular
T  =  Triangular
B  =  Blackman
H  =  Hamming
A  =  Hanning
```

For a standard periodogram, rectangular window is used, and it is the default window type. We also need to specify whether the periodogram is to be computed on a linear scale or a logarithmic (decibel) scale.

```
PSD scaling:  2      1  =  Linear
                     2  =  Logarithmic
```

Finally, we need to specify the name of the variable that will hold the data points for the periodogram to be computed. The default variable name is TEMPR.

```
Variable name for periodogram :    TEMPR
```

Accept the default variable name. After a few seconds, the following two lines will be displayed:

```
Variable = TEMPR     , Type = R , Size =  256
Variable = NFREQ     , Type = R , Size =  256
```

Similar to filter analysis routines covered in Chapters 6 and 7, the second variable NFREQ holds the normalized frequencies that correspond to the samples in TEMPR. A quick-plot can be obtained at this point by pressing the [F2] function key (see Fig. 8-2).

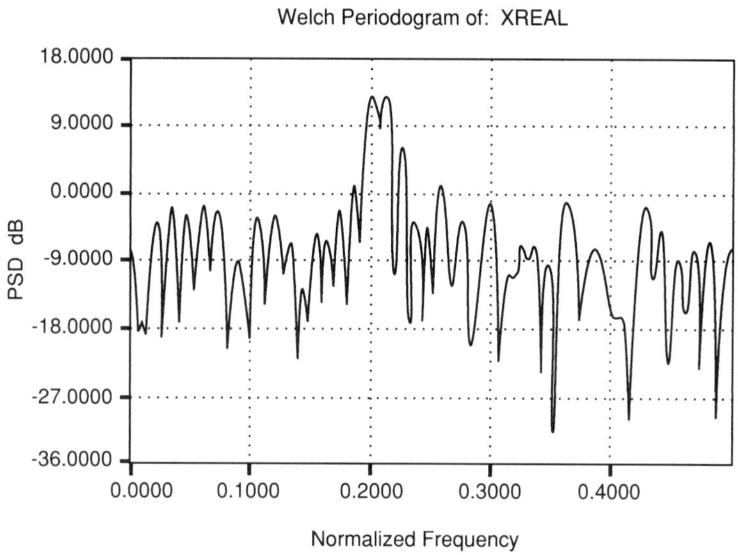

Figure 8-2. Periodogram for the sequence XREAL.

Example 8-2:

Repeat Ex. 8-1, this time using the complex data sequence XCOMP in the data file EX8-2.DSP. Select *Periodogram*, and make the same entries as in the previous example. The resulting periodogram is shown in Fig. 8-3. There is one major difference between the two periodograms computed in Exs. 8-1 and 8-2. In Ex. 8-1, 256 samples of the periodogram were computed for the range of normalized frequencies from 0.0 to 0.5. In Ex. 8-2, 512 samples are computed, so that the frequency axis extends from 0.0 to 1.0. Recall that for a real sequence, the second half of the periodogram is a mirror image of the first half and is therefore redundant. For a complex sequence, however, the second half of the periodogram is no longer a mirror image of the first half and must be shown.

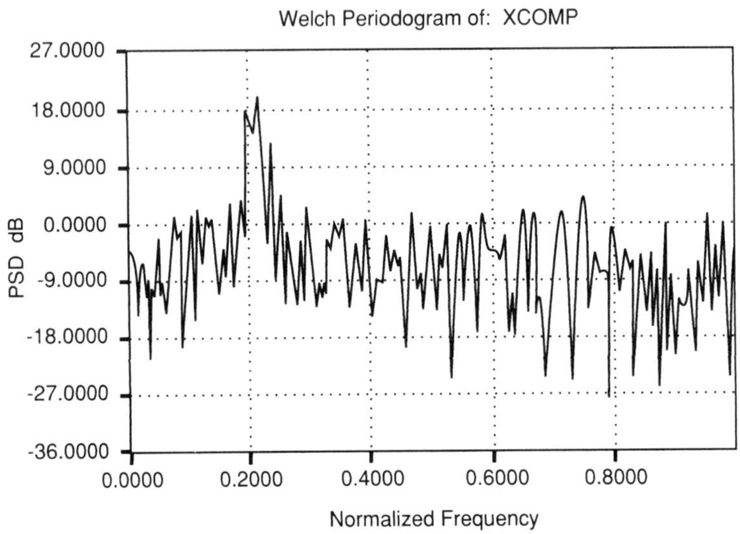

Figure 8-3. Periodogram for the sequence XCOMP.

Example 8-3

Compute and plot the Welch periodogram for the sequence XREAL using a Hamming window and 30-sample segments that overlap by 5 samples.

Select *Periodogram*, and enter the following parameter values:

```
Variable name for input data  :    XREAL

Segment length :    30          Segment overlap :    5

Window type    :    H          PSD scaling     :    2

Variable name for periodogram :    TEMPR
```

The resulting PSD estimate is shown in Fig. 8-4. Notice the reduction in the side lobes and the broadening of the main lobe around the normalized frequency 0.2. This is mainly due to the window used. Also notice the reduction in the variability of the PSD estimate, which is a result of averaging periodograms.

8-2 BLACKMAN-TUKEY SPECTRAL ESTIMATE

The Blackman-Tukey estimation method consists of three steps. In the first step, the middle $2M+1$ samples of the autocorrelation sequence, $\phi_{xx}(m), -M \leq m \leq M$, are estimated from the available N-point data record. The second step is to apply a data

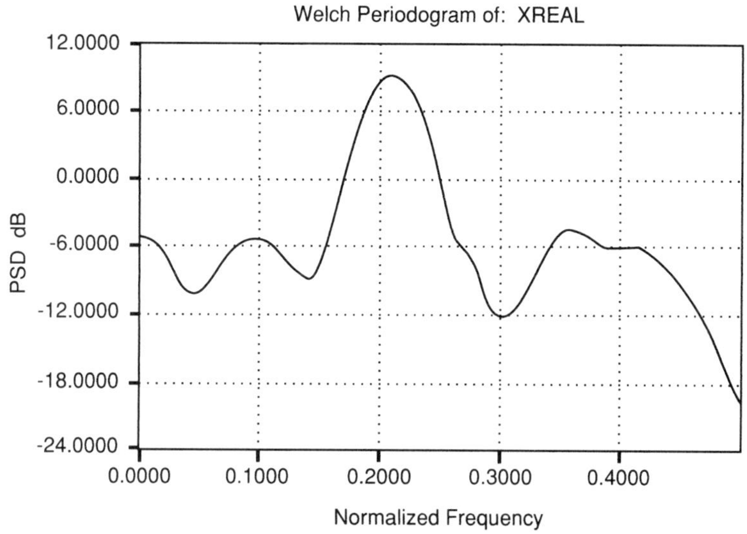

Figure 8-4. PSD estimate for XREAL, computed in Ex. 8-3.

window to the estimated autocorrelation lags. Finally, the windowed autocorrelation estimate is Fourier-transformed to yield the Blackman-Tukey PSD estimate. Mathematical formulation of this procedure can be given with the following equations:

$$r_{xx}(m) = \frac{1}{N} \sum_{N} x^*(n)x(n+m), \qquad -M \le m \le M$$

$$\tilde{r}_{xx}(m) = w(m)r_{xx}(m)$$

$$P_{BT}(F) = \sum_{m=-M}^{M} \tilde{r}_{xx}(m) \exp(-j2\pi Fm) \tag{8-2}$$

For some data records and windows, the PSD estimate may contain negative values. Among the window functions available in PC-DSP, only the triangular window guarantees a nonnegative estimate.

Example 8-4

Compute and plot the Blackman-Tukey PSD estimate (in decibels) for the sequence XREAL. Use $M = 30$ with a triangular window.

Select *Spectrum*, and then select the *Black-Tukey* function. Enter the following parameter values in response to the prompts.

```
Variable name for input data  :   XREAL
Largest autocorrelation lag   :   30
```

```
Window type    :    T          PSD scaling    :    2

Variable name for periodogram :    TEMPR
```

As in the periodogram estimate, two variables are generated with the names TEMPR, and NFREQ. The *Blackman-Tukey* PSD estimate can be plotted by pressing the [F2] key and is shown in Fig. 8-5.

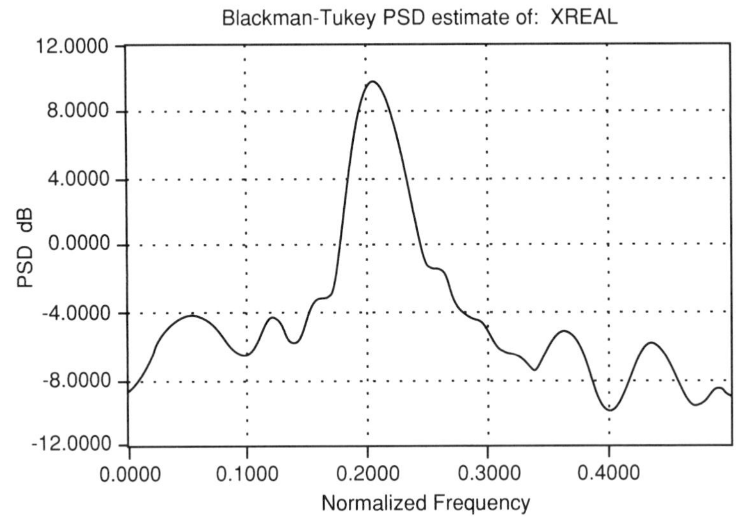

Figure 8-5. Blackman-Tukey PSD estimate for XREAL.

8-3 AUTOREGRESSIVE SPECTRAL ESTIMATION

We will begin by presenting some background information on autoregressive (AR) modeling and spectral estimation. Keep in mind that mathematical rigor will be sacrificed for the sake of simplicity.

In AR modeling of data, the observed data record is assumed to have been generated by driving an all-pole linear system with white noise. An all-pole system with M poles is characterized by a transfer function in the form

$$H(z) = \frac{1}{A(z)} \tag{8-3}$$

where

$$\begin{aligned} A(z) &= 1 + a_1 z^{-1} + a_2 z^{-2} + \cdots + a_M z^{-M} \\ &= 1 + \sum_{k=1}^{M} a_k z^{-k} \end{aligned}$$

If the input to this system is a white noise process $w(n)$ with zero mean and variance σ^2, the power spectral density of the resulting output process $x(n)$ is

$$\begin{aligned} P_x(F) &= |H(e^{j2\pi F})|^2 P_w(F) \\ &= |H(e^{j2\pi F})|2\sigma^2 \end{aligned} \tag{8-4}$$

and the difference equation relating the input and the output signals of the all-pole filter can be written as

$$x(n) = -\sum_{k=1}^{M} a_k x(n-k) + w(n) \tag{8-5}$$

which leads to a linear prediction interpretation of AR modeling. The coefficients $a_1, a_2, \ldots, a_M$ are the coefficients of the Mth-order AR model for $x(n)$. The last coefficient in the set, i.e., a_M, is called a reflection coefficient. The AR spectrum estimation problem can be stated as follows: Given the data record $x(n)$ and the model order M, estimate the set of AR coefficients $a_1, a_2, \ldots, a_M$, and the driving noise variance σ^2. Then, use Eq. (8-4) to compute the PSD estimate.

PC-DSP uses two methods for estimation of AR coefficients: (1) the Yule-Walker method and (2) the Burg method. Using either method, any of the following can be computed: (1) PSD estimate for the range of normalized frequencies from 0.0 to 0.5 for real data or from 0.0 to 1.0 for complex data, (2) AR coefficients for a specified model order, (3) reflection coefficients up to a specified model order, and (4) driving noise variance estimates up to a specified model order.

Example 8-5

Using the Yule-Walker method, compute coefficients of the 30th-order AR model for the sequence XREAL. Also compute and plot the dB PSD estimate.

Select *Spectrum* and then *Auto-Regr*. Answer prompts as follows:

```
Variable name for input data  :   XREAL

Analysis method :    Y           AR model order :    30

Analysis type   :    A

Variable name for results :    ARCOEFF
```

Notice that possible answers to the analysis method question are

B = Burg's ME algorithm

Y = Yule – Walker algorithm

and available analysis types are

P = PSD estimate

A = AR model coefficients

R = reflection coefficients

V = driving noise variance

The coefficients a_1 thru a_{30} are now stored in the variable ARCOEFF. To compute the PSD estimate, we select the *Auto-Regr* function again and make the same entries, with the exception of the analysis type, which will be P. Also, the PSD scaling needs to be specified as logarithmic. (Note that this second part of the problem is independent of the first part. We could have done this without first storing the AR coefficients in the variable ARCOEFF.) After the PSD estimate is computed, two variables with the names TEMPR (default) and NFREQ are generated by the program. The PSD estimate can now be plotted by pressing the [F2] key (see Fig. 8-6).

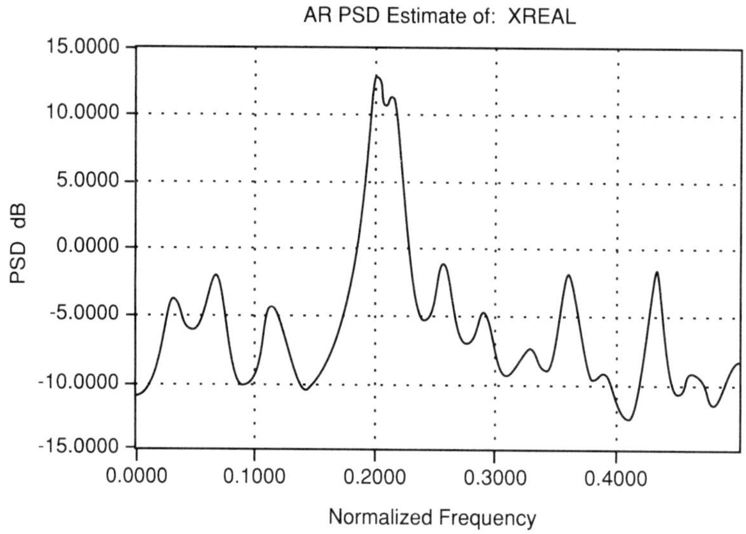

Figure 8-6. AR PSD estimate of XREAL (Yule-Walker method, $M = 30$).

Example 8-6

An important problem in the AR spectral analysis is the estimation of the best model order for a given signal. If the model order chosen is too low, an oversmoothed PSD estimate is obtained. If the model order is too high, the PSD estimate shows spurious detail. One widely used criterion for estimating the model order is the *Akaike Information Criterion* (AIC) given by

$$\text{AIC}(M) = \ln(\sigma_M^2) + \frac{2M}{N} \tag{8-6}$$

The model order is chosen to minimize AIC(M). In this example, we will generate the sequence AIC(M) for the data record in XREAL.

Select *Auto-Regr* again and compute the driving noise variances of models up to 30th-order with the following entries:

```
Variable name for input data  :    XREAL

Analysis method :    Y          AR model order :    30

Analysis type   :    V

Variable name for results :    SIGMA2
```

The sequence AIC(M) can be generated by using the available functions in the *Data* component. The minimum value is obtained for $M = 5$ (see Fig. 8-7).

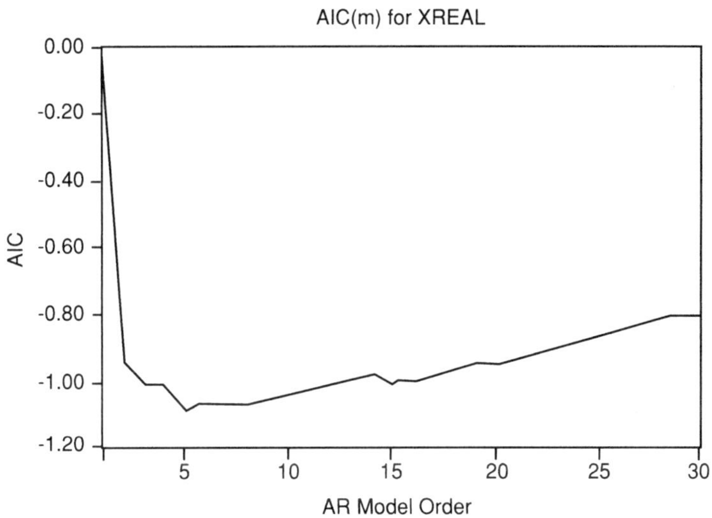

Figure 8-7. Akaike information criterion for XREAL (Yule-Walker method).

8-4 MOVING AVERAGE SPECTRAL ESTIMATION

In contrast to AR modeling, a moving-average (MA) model is an all-zero model, where the process $x(n)$ is assumed to have been generated by driving a FIR filter with white noise, i.e.,

$$x(n) = w(n) + \sum_{k=1}^{M} b_k w(n-k) \tag{8.7}$$

The corresponding system function is

$$H(z) = 1 + \sum_{k=1}^{M} b_k z^{-k}$$

If the coefficients $b_1, b_2, \ldots, b_k$ are known, the PSD of the output process can be computed as

$$P_x(F) = |1 + \sum_{k=1}^{M} b_k \exp(-j2\pi Fk)|^2 \sigma^2 \tag{8.8}$$

PC-DSP uses Durbin's method to estimate the coefficients of the MA model of a process from which an N-point data record is available. This method consists of two steps. In the first step, the Yule-Walker method is used to model the input sequence as an Lth order AR process, where $M << L << N$. In the second step, the coefficients of this Lth order AR model are used as input data to compute a set of M coefficients $b_1, \ldots, b_M$, again using the Yule-Walker method.

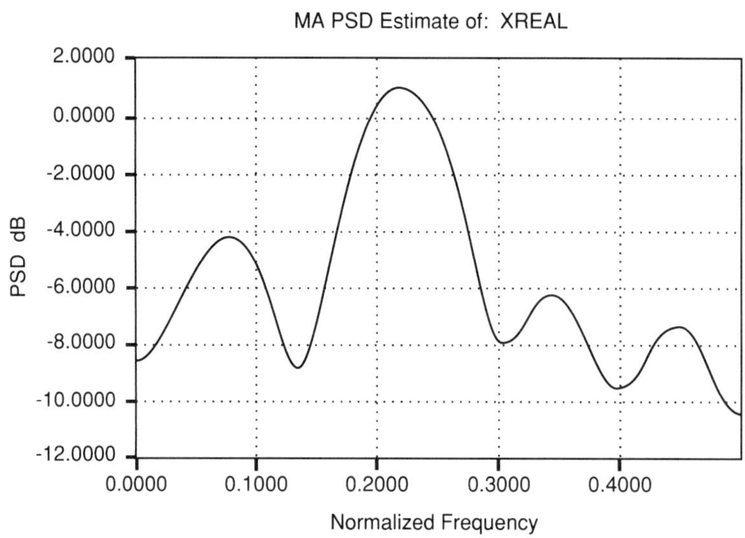

Figure 8-8. MA PSD estimate for XREAL.

Example 8-7

Compute and plot the tenth-order MA PSD estimate for the sequence XREAL. Use 16 coefficients for the large-order AR model.

Select *Moving-Avg*, and make the following entries:

```
Variable name for input data  :    XREAL

MA model order :    10           AR model order :    16
```

```
Analysis type  :  P          PSD scaling  :  2

Variable name for results :  TEMPR
```

Coefficients of the MA model can also be stored in a named variable for tabulation by answering the analysis type prompt with an M instead of a P.

8-5 ARMA SPECTRAL ESTIMATION

An autoregressive moving average (ARMA) process is generated by driving a pole-zero filter with white noise. The z-domain system function of the filter is in the form

$$H(z) = \frac{B(z)}{A(z)} \tag{8.9}$$

where

$$A(z) = 1 + \sum_{k=1}^{Q} a_k z^{-k}$$

$$B(z) = 1 + \sum_{k=1}^{P} b_k z^{-k} \tag{8.10}$$

If the numerator and the denominator coefficients are known, the PSD of the output process can be computed as

$$P_x(F) = |H(e^{j2\pi F})|2\sigma^2$$

Finding an optimum ARMA model estimate for a given process is computationally quite difficult, since the equations to be solved are nonlinear. PC-DSP uses an ad hoc method for finding a "nearly optimum" solution. First, the AR coefficients $a_1, a_2, \ldots, a_Q$ are estimated using the Yule-Walker method. The sequence $x(n)$ is then filtered through a FIR filter with the system function $A(z)$ given by Eq. (8.10). Finally, the MA coefficients $b_1, b_2, ..., b_P$ are estimated from the output sequence of the filter using the Durbin method.

Example 8-8

Compute and plot the ARMA(10,10) PSD estimate for the sequence XREAL.
Select *aRma* and make the following entries:

```
Variable name for input data  :  XREAL

MA model order :  10          AR model order :  10
```

```
Analysis type  :    P              PSD scaling    :   2

Variable name for PSD estimate :    TEMPR
```

Figure 8-9 shows the PSD estimate obtained.

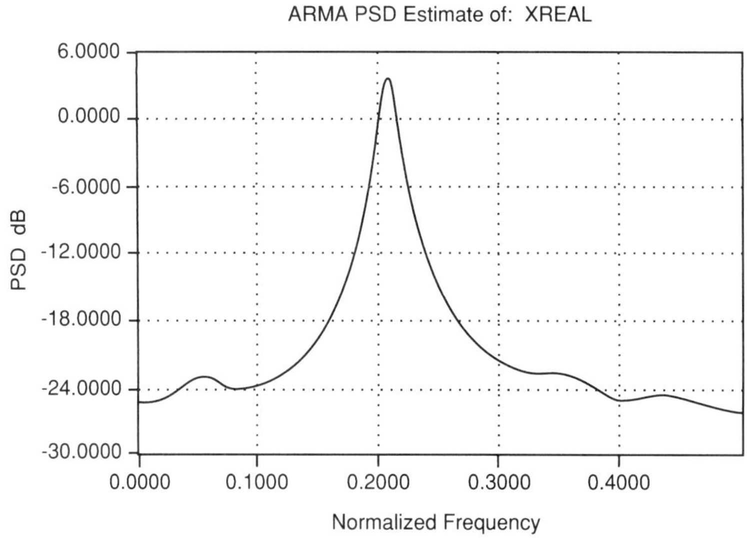

Figure 8-9. ARMA PSD estimate for XREAL.

8-6 EXERCISES

1. The purpose of this exercise is to study the variance of the standard periodogram and to see its inconsistency as an estimator. We will also see how averaging periodograms affects the variance of the estimate in the Welch method.

a. Generate a 100-point Gaussian random sequence with zero mean and unit variance. Compute and plot the decibel periodogram. How does your periodogram compare with the true PSD of Gaussian white noise? Comment on the variability of the PSD estimate.

b. Repeat part (a) with a 200-point Gaussian random sequence. Does the increase in the number of samples result in a decrease in the estimator variance?

c. Repeat part (a) with a 1000-point random sequence. Does this improve the PSD estimate?

d. Estimate the PSD for the 1000-point sequence of part (c) using the Welch method. Use a segment length of 100, with no overlapping samples between segments. Choose the rectangular window. Compare the variability of the PSD estimate to that of part (c).

e. Repeat part (d) with a triangular window. How does the PSD estimate differ from the estimate obtained in part (d)?

2. The purpose of this exercise is to study the frequency separation capability of the periodogram for sinusoids that are closely spaced in frequency.

a. Generate 30 samples of the following signal:

$$x(n) = \sin[2\pi(0.2)n] + \sin[2\pi(0.22)n], \qquad n = 0, \ldots, 29$$

Compute and plot the standard decibel periodogram for $x(n)$. Does the periodogram show two sinusoids?

b. Generate 100 samples of the sequence given in part (a) repeat the same problem. Does the periodogram indicate two sinusoids now? Explain.

c. Repeat part (b) using a triangular window. How does the window affect frequency separation? Explain.

3. In this exercise, the use of data windowing to reduce spectral leakage will be demonstrated.

a. Generate the following 100-sample sequence

$$x(n) = 2\sin[2\pi(0.2)n] + 0.1\sin[2\pi(0.24)n], \qquad n = 0, \ldots, 99$$

b. Sketch the true power spectral density of $x(n)$.

c. Compute and plot the periodogram for $x(n)$. Does the periodogram show both sinusoids? If not, why?

d. Repeat part (c), this time using a Hamming window. How does the window affect the detection of the weaker sinusoid?

4. In this exercise, we will examine the effects of varying the signal to noise ratio (SNR) on the AR PSD estimates. We will use noisy sinusoids in the form

$$x(n) = A\sin(2\pi Fn) + w(n) \qquad F = 0.2$$

where $w(n)$ is a white noise process with zero mean. The signal to noise ratio is $A^2/2\sigma^2$, where σ^2 is the variance of the additive noise process.

a. Generate 30 points of $x(n)$ with SNR = 30 dB. Compute and plot the third-order AR PSD estimate using the Yule-Walker method.

b. Repeat part (a) for SNR values 20, 10, 5, 0 dB. How does changing the SNR affect the sharpness of the peak ?

c. Repeat parts (a) and (b) with the Burg method.

5. Use the sequence $x(n)$ obtained in Exercise 4 with SNR = 10 dB. In an attempt to increase the sharpness of the peak, the AR model order is to be increased.

a. Using the Burg method, compute and plot AR PSD estimates for model orders $M = 6, 9, 12$.

b. Repeat (a) using the Yule-Walker method.

6. We will use the 30-sample sequence $x(n)$ that was generated in Exercise 2.

a. Using the Yule-walker method, compute and plot the fifth-order AR PSD estimate. Are the two peaks resolved?

b. Repeat part (a) using the Burg method. Are the peaks resolved now?

c. Generate a 30-sample sequence $w(n)$ from a white noise process with mean = 0 and variance = 0.2. Add $w(n)$ to the sequence $x(n)$. Repeat part (b) with this noisy sequence. How does the additive noise affect the resolution?

d. Repeat part (a) for model orders 10 and 20.

9

Advanced Applications

Another powerful feature of PC-DSP is its expansion capability. Using the macro compiler DSPMAC that is supplied on the PC-DSP Program Disk, a set of PC-DSP commands can be compiled into a macro file that is executed automatically from the *Macro* component. This feature makes it possible for the user to add new functions to PC-DSP by combining the existing functions in a certain way. Also, by using disk file input-output functions and the *OS-shell* function in a macro file, it is possible to interface stand-alone executable programs to PC-DSP and make them part of an integrated environment. Procedures for expanding the capabilities of PC-DSP are discussed in this chapter.

9-1 PC-DSP MACRO LANGUAGE

The macro language of PC-DSP implements all the functions that are found in the *Data* component of the program. Additionally, *Load,* *Save*, and *OS-shell* functions of the *System* component are implemented. To the extent possible, the syntax of the macro language is kept consistent with the entry of parameters in the interactive mode. As an example, consider the following problem to be solved using PC-DSP:

1. Generate a 10-sample ramp sequence XN with initial value 0 and increment 0.1.
2. Compute the DFT of the sequence XN and put the result in XK.

3. Compute the magnitude of XK and put the result in XM.

To solve this problem in interactive mode, we would perform the following steps:

1. Select *Ramp*. Type:

```
XN <ENTER> 0 <ENTER> 0.1 <ENTER> 100 <ENTER>
```

2. Select *DFT*. Type:

```
XK <ENTER> XN <ENTER>
```

3. Select *Magnitude*. Type:

```
XM <ENTER> XK <ENTER>
```

To automate this procedure, a macro file can be written with the following three lines in it.

```
RAMP  XN  0  0.1  100
DFT  XK  XN
MAGNITUDE  XM  XK
```

Each line contains the name of the function to be used, followed by the parameters for that function, in the same order that they would be entered in interactive mode. Table 9-1 lists the functions that have syntax identical to their respective counter parts in the interactive mode. Functions that are unique to the macro language are listed in Table 9-2.

The usage of the functions in Table 9-1 has been previously discussed in Chapters 3 and 4. We will discuss the functions in Table 9-2 in detail. For now, let us continue with the simple three-line macro file example. Assuming that this file is created using an ASCII text editor and saved to disk under the name MYMACRO.SRC, we will invoke the macro compiler DSPMAC to compile it. At the operating system prompt, type

```
DSPMAC    <ENTER>
```

We will compile MYMACRO.SRC to obtain MYMACRO.MAC. Answer the compiler prompts as follows:

```
Source file [.SRC] : MYMACRO
Output file [.MAC] : MYMACRO
```

The compiler will respond with the message

TABLE 9-1 PC-DSP MACRO LANGUAGE FUNCTIONS

Function	Parameters			
CONJG	Var1	Var2		
COSINE[a]	Var1	Var2		
EVEN	Var1	Var2		
EXP[a]	Var1	Var2		
FLIP	Var1	Var2		
IMAG	Var1	Var2		
LOGARITHM	Var1	Var2		
MAGNITUDE	Var1	Var2		
ODD	Var1	Var2		
PHASE	Var1	Var2		
REAL	Var1	Var2		
RECIPR	Var1	Var2		
SINE[a]	Var1	Var2		
SQ-ROOT	Var1	Var2		
TANGENT	Var1	Var2		
FFT	Var1	Var2		
INV-FFT	Var1	Var2		
DFT	Var1	Var2		
INV-DFT	Var1	Var2		
ADDS[b]	Var1	Var2	Var3	
MULTIPLYS[b]	Var1	Var2	Var3	
CONVOLVE	Var1	Var2	Var3	
CROSS-CORR	Var1	Var2	Var3	
JOIN	Var1	Var2	Var3	
MAKECOMP	Var1	Var2	Var3	
DECIMATE	Var1	Var2	Factor	
SHIFT	Var1	Var2	#Sampl	
ZERO-INS	Var1	Var2	Factor	
ZERO-PAD	Var1	Var2	#Sampl	
IMPULSE	Var1	#Sampl		
RECTANGULAR	Var1	#Sampl		
TRIANGULAR	Var1	#Sampl		
BLACKMAN	Var1	#Sampl		
HAMMING	Var1	#Sampl		
HANNING	Var1	#Sampl		
ADDC[c]	Var1	Var2	Const	
MULTIPLYC[c]	Var1	Var2	Const	
RAMP	Var1	Init	Incr	#Sampl
NORMAL	Var1	Mean	Variance	#Sampl
UNIFORM	Var1	Lower	Upper	#Sampl

[a] Equivalent to its counterpart in *Nonlinear* submenu.
[b] Add/multiply sequences sample by sample.
[c] Constant add/multiply.

```
Lines = 3     Errors = 0
```

At this point, a file with the name MYMACRO.MAC should exist on the default disk drive. This macro can be executed by selecting the *Macro* component from the main

TABLE 9-2 FUNCTIONS UNIQUE TO MACRO LANGUAGE

Function	Parameters			
ALOAD	Filename	Varname	Vartype	
ASAVE	Varname	Filename		
DLOAD	Filename			
DSAVE	Varname	Filename		
WINDOW	Row1	Column1	Row2	Column2
GETSTR	Row	Column	Prompt	Regnum
GETFLP	Row	Column	Prompt	Regnum
GETINT	Row	Column	Prompt	Regnum
OS-SHELL	Command			
PARFILE	Filename			
MESSAGE	String			

menu of PC-DSP. When prompted, type the name of the macro to be executed.

```
File name :     MYMACRO
```

After execution of the macro is completed, the *Var-list* function can be used to verify that sequences XN, XK, and XM are stored in memory. Also, the last generated sequence, XM, can be graphed by pressing the [F2] function key.

9-2 SCREEN I/O USING REGISTERS

The macro file MYMACRO.MAC that we created in the last section performs the same sequence of functions with the same parameters every time it is executed. If we wanted to change a parameter, say the increment of the ramp sequence XN, we would have to edit the source file and recompile it. The macro file would certainly be more useful if the user had some way of specifying parameter values at execution time. This can be achieved using the macro registers of PC-DSP.

PC-DSP has a total of 15 internal registers that can be accessed by macro files. Registers are identified by the @ character followed by one of the letters S, F, or I and then a single digit from 1 to 5. The letter immediately following the @ character indicates the type of the register: S = string register, F = floating-point register, I = integer register. The single digit following the register type is the register number. There are five registers of each type. As an example, @F2 indicates the second floating-point register.

Registers can be used in place of formal parameters in macro files. Consider the first line in the file MYMACRO. To make the increment of the ramp sequence a variable instead of fixing its value in the macro file, we can modify the first line of the file MYMACRO.SRC as follows:

```
RAMP  XN  0  @I1  100
```

Now the increment is taken from the integer register 1 at the time this statement is executed. This makes it possible to change the value of the increment without having to change the macro file. We simply load the register with a different value and run the macro again.

There are three screen I/O functions that are used for loading registers with values entered by the user through the keyboard. These functions are GETSTR, GETFLP, and GETINT. Consider the following statement:

```
GETSTR  10  15  Enter_the_variable_name_:__  2
```

When executed in a macro file, this statement causes the prompt *"Enter the variable name:"* to be printed on the screen beginning at the tenth row and fifteenth column position. Underscore characters in the prompt string are not printed on the screen, they simply indicate blank spaces between words. After printing this prompt, PC-DSP waits for the user to respond. When finished, the character string typed by the user is stored in the string register 2 and is available for use by other statements in the macro file. Similarly, GETFLP is used for screen entry and storage of floating-point numbers, and GETINT is used for integers.

9-3 A COMPLETE MACRO EXAMPLE

As an exercise, we will develop a macro file for computing the circular convolution of two real sequences. Our macro file will be named CIRCONV and will perform the following steps:

1. Print a brief description of this macro on the screen. We will use the functions WINDOW and MESSAGE for this purpose.

```
WINDOW  4  35  10  78
MESSAGE  6  38  This_function_computes_the_circular
MESSAGE  7  38  convolution_of_two_sequences.
```

The WINDOW statement opens a window with the upper left corner at row-column position 4-35, and the lower right corner at row-column position 10-78. The MESSAGE statements that follow print the messages at indicated row-column positions.

2. Prompt the user for the following parameters and store each parameter in the register indicated.

$$Output\ sequence \longrightarrow string\ register\ 1$$

$$Input\ sequence\ 1 \longrightarrow string\ register\ 2$$

$$Input\ sequence\ 2 \longrightarrow string\ register\ 3$$

$$Number\ of\ samples\ on\ the\ circle \longrightarrow integer\ register\ 1$$

```
WINDOW  12  3  22  78
GETSTR  14  6  Var1_:__   1
GETSTR  16  6  Var2_:__   2
GETSTR  18  6  Var3_:__   3
GETINT  20  6 #Sampl_:__   1
```

3. Since the DFT method will be used for computing the circular convolution, we have to ensure that both inputs have the same number of samples, which is stored in integer register 1. We will pad input sequences with zeros if necessary. This part is tricky! The ZERO-PAD function cannot be used, since we don't know ahead of time how many zeros to append. A possible solution is to generate an N-point sequence with all samples equal to zero and then add this sequence to both inputs. (Recall that, in sample by sample addition of two sequences with unequal lengths, the length of the result is equal to the length of the longer sequence.) For this method to work, however, the user must ensure that neither input sequence is longer than N points. The RAMP function will be used with zero initial value and zero increment for generating the sequence of N zeros.

```
RAMP  TEMPR  0  0  @I1
ADDS  SEQ2  @S2  TEMPR
ADDS  SEQ3  @S3  TEMPR
```

4. Now the variables SEQ2 and SEQ3 contain the input sequences extended to length N. In the last step, we will compute DFTs of SEQ2 and SEQ3, multiply them, and take the inverse DFT of the product. Finally, we will take the real part of the result to obtain the desired circular convolution. We are assuming that this function will be used with real input sequences, in which case the output must also be real. However, because of machine round-off error in the DFT routine, the imaginary part of the output sequence will be nonzero, although very small. It is necessary to take the real part for the quick-plot to work.

```
DFT  SEQ2  SEQ2
DFT  SEQ3  SEQ3
MULTIPLYS  @S1  SEQ2  SEQ3
INV-DFT  @S1  @S1
REAL  @S1  @S1
```

This completes the development of CIRCONV. A complete listing of the source code is given in Fig. 9-1. The source file can be compiled using DSPMAC and called from the Macro component of PC-DSP.

9-4 INTERFACING USER-WRITTEN PROGRAMS TO PC-DSP

The macro language of PC-DSP also provides the capability to send a sequence to an external program (referred to as a *child process*), process it outside PC-DSP, and bring the

```
& Circular convolution routine for PC-DSP.
&
& Oktay Alkin, 01/23/1988.
&
WINDOW   4  35  10  78
MESSAGE  6  38  This_function_computes_the_circular
MESSAGE  7  38  convolution_of_two_sequences.
WINDOW  12   3  22  78
&
& Var2 and Var3 are circularly convolved, and the
& result is stored in Varl.  #Sampl is the number
& of points on the circle.
&
GETSTR  14  6  Varl_:__   1
GETSTR  16  6  Var2_:__   2
GETSTR  18  6  Var3_:__   3
GETINT  20  6  #Sampl_:__   1
&
& Generate a sequence TEMPR with #Sampl zeros.
& Add TEMPR to Var2 and Var3 to obtain sequences
& SEQ1 and SEQ2.
RAMP   TEMPR  0   0  @I1
ADDS   SEQ2  @S2  TEMPR
ADDS   SEQ3  @S3  TEMPR
&
& Compute DFT's of SEQ2 and SEQ3, then multiply and invert.
&
DFT  SEQ2  SEQ2
DFT  SEQ3  SEQ3
MULTIPLYS  @S1  SEQ2  SEQ3
INV-DFT  @S1  @S1
&
& Remove imag. part due to round-off error.
& Tell the user we are finished.
&
REAL  @S1  @S1
MESSAGE  9  38  Done.
```

Figure 9-1. Source code for the macro file CIRCONV.

results back into PC-DSP for graphing or further processing. Furthermore, this procedure can be automated and made transparent to the user.

The transfer of data sequences from PC-DSP to the child process and back is achieved through the use of ASAVE and ALOAD functions in a macro file. The ASAVE function saves a PC-DSP variable into an ASCII file which may be accessed by the child process. In contrast, the ALOAD function loads a variable from an ASCII file (which may have been written by the child process).

A child process is an executable file (typically a DOS file with the extension COM or EXE) that is written in any programming language (such as FORTRAN, PASCAL, or BASIC), and compiled. The OS-SHELL statement of the macro language is used for invoking a child process. When used in a macro file, OS-SHELL differs from its interactive-mode counterpart. It accepts one argument, which is the DOS command necessary to invoke the child process. As an example, consider a macro file with the following three lines:

```
ASAVE   XN   TRANSFER.DAT
OS-SHELL   MYPROG
ALOAD   TRANSFER.DAT   YN   R
```

The first line of the macro file causes the sequence XN to be saved into the ASCII file TRANSFER.DAT. The second line invokes the user program MYPROG.EXE as a child process. Presumably, the child process reads the sequence in the data file TRANSFER.DAT, processes it, and saves the resulting sequence back into the same data file, overwriting its original contents. As soon as the child process terminates, control is passed to the next statement in the macro file, which loads the contents of TRANSFER.DAT into PC-DSP as a real sequence with the name YN.

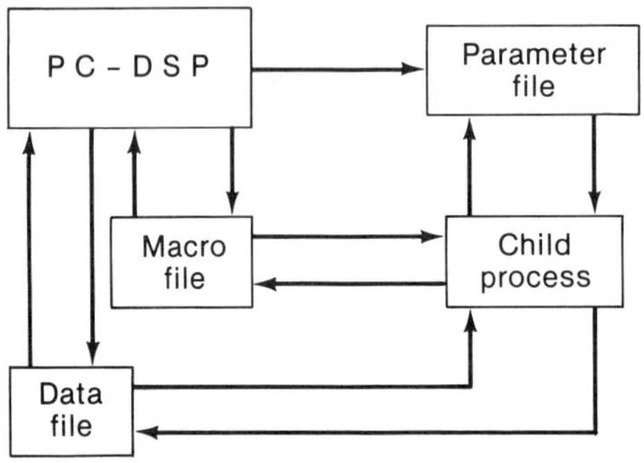

Figure 9-2. Interfacing a child process to PC-DSP.

Typically, the first task of the child process is to read the contents of an ASCII data file created by PC-DSP just before calling the child process. In most programming languages, reading and writing ASCII data files is quite simple and straightforward. There is one important detail that requires special attention in reading a data file written by PC-DSP: An ASCII data file should be read *till the end*, i.e., until an end-of-file marker is reached. The number of samples to be read is usually not known in advance, and there is no information in the file about this number. Data file read-write routines and sample child processes in FORTRAN, PASCAL, and compiled BASIC languages are provided next. These routines can be modified to suit particular needs.

FORTRAN: Subroutines DATAIN and DATAOU can be used in a child process for sequential (ASCII) data file input and output. They were tested using the WATFOR-77 FORTRAN compiler available from WATCOM Inc. and should also work on any other FORTRAN compiler that conforms to the 1977 ANSI standards for FORTRAN compilers.

```
      SUBROUTINE DATAIN(X, NPTS)
C
C   Reads the sequence in the data file TRANSFER.DAT .
C   Upon return, the array X contains the sequence, and
C   the integer variable NPTS contains the number of
C   samples read.
C
      DIMENSION X(1)
      OPEN(UNIT=1, FILE='TRANSFER.DAT', STATUS='OLD')
      I = 0
   10 I = I + 1
      READ(1, *, END=20) X(I)
      GOTO 10
   20 NPTS = I - 1
```

```
      CLOSE(1)
      RETURN
      END
      SUBROUTINE DATAOU(X ,NPTS )
C
C   Writes NPTS samples in the array X into the data
C   file TRANSFER.DAT .
C
       DIMENSION X(1)
       OPEN(UNIT=1, FILE='TRANSFER.DAT')
       DO 10 I=1, NPTS
       WRITE(1, *) X(I)
  10   CONTINUE
       RETURN
       END
```

The main body of the FORTRAN program should first dimension the arrays to be used. It should then read the input data by calling DATAIN, perform calculations, and write the output data to disk by calling DATAOU. A sample main program is as follows:

```
C   Sample child process for PC-DSP. This sample
C   program rectifies a given sequence.  It can be
C   modified to suit user's needs.
C
      DIMENSION X(4096)
      CALL DATAIN(X, NPTS)
C
      DO 10 I=1, NPTS
      IF(X(I).LT.0.)    X(I) = 0.
  10  CONTINUE
C
      CALL DATAOU(X, NPTS)
      STOP
      END
```

PASCAL: Source listings are given next for procedures DataIn and DataOu. These procedures were tested using the Turbo PASCAL compiler version 3.0 available from Borland International and may require slight modifications if they are to be compiled on other PASCAL compilers.

```
PROCEDURE DataIn ( var X:    Sequence;
                   var Npts:  Integer   );
VAR
  Count:  Integer;
  DataFile:  Text;
  BEGIN
  ASSIGN (DataFile, 'TRANSFER.DAT');
```

```
  RESET (DataFile);
  Count := 0;
  WHILE NOT EOF (DataFile) DO
    BEGIN
      Count := Count + 1;
      READLN (DataFile, X[Count]);
    END;
  Npts := Count;
  CLOSE (DataFile)
END;
PROCEDURE DataOu ( var X:    Sequence;
                   var Npts:  Integer   );
VAR
  DataFile:  Text;
  i:  Integer;
BEGIN
  ASSIGN (DataFile, 'TRANSFER.DAT');
  REWRITE (DataFile);
  FOR i := 1 to Npts do
    WRITELN (DataFile, X[i]);
  CLOSE (DataFile)
END;
```

Using procedures DataIn and DataOu, the rectifier program can be written in PASCAL as follows:

```
(* Sample child process for PC-DSP. This sample
   program rectifies a given sequence.  It can be
   modified to suit user's needs.
*)
   PROGRAM Child_Process;
   TYPE
     Sequence = Array [1..4096] of Real;
   VAR
     X: Sequence;
     i, Npts:  Integer;
   {$I datain.pas}     (* Include file i/o routines *)
   {$I dataou.pas}     (*    COMPILER DEPENDENT !!!  *)
   BEGIN
     DataIn (X, Npts);
     FOR i := 1 TO Npts DO
       IF X[i] < 0.0 THEN X[i] := 0.0;
     DataOu (X, Npts)
END.
```

Compiled BASIC: In recent years, compiled BASIC has gained a great deal of popularity. This is due to several factors: Line numbers are not required. As in FORTRAN, subroutines may be separated from the main body of the program and may have their local variables. Compiled BASIC versions of the data I/O routines DataIn and DataOu are given below. These routines were tested using the Microsoft QuickBASIC compiler version 4.5.

```
SUB DataIn (X(), Npts%) STATIC
  OPEN "I", #1, "TRANSFER.DAT"
  Count% = 0
  WHILE NOT EOF(1)
    Count% = Count% + 1
    INPUT #1, X(Count%)
  WEND
  Npts% = Count%
  CLOSE #1
END SUB
SUB DataOu (X(), Npts%) STATIC
  OPEN "O", #1, "TRANSFER.DAT"
  FOR Count% = 1 TO Npts%
    PRINT #1, X(Count%)
  NEXT Count%
  CLOSE #1
END SUB
```

Source listing for the compiled BASIC version of the rectifier program is as follows:

```
Rem...  Sample child process for PC-DSP. This sample
Rem...  program rectifies a given sequence.  It can be
Rem...  modified to suit user's needs.
Rem...
  DIM X(4096)
  CALL DataIn(X(), Npts%)
  FOR Count% = 1 TO Npts%
    IF X(Count%) < 0 THEN X(Count%) = 0
  NEXT Count%
  CALL DataOu(X(), Npts%)
END
```

One final note on ASCII data file I/O between PC-DSP and child processes: The environment variable TEMP can be used to force PC-DSP into using another subdirectory for temporary data file I/O. If a statement such as

```
SET TEMP=C:\TEMPFLS\
```

is used before invoking PC-DSP, all temporary files that are used in conjunction with macro language functions ALOAD, ASAVE, DLOAD, DSAVE, and PARFILE are written into and read from the TEMPFLS subdirectory of the C: drive. Of course, the child process must be aware of this and can check the same environment variable if necessary.

If a RAM disk is available, it can be specified as the temporary directory to speed up data transfer between PC-DSP and child processes.

9-5 AN EXAMPLE: LATTICE FILTER

In this section, we will present a complete example of interfacing a FORTRAN program to PC-DSP. The program under consideration is a FIR filter implemented as a lattice structure. Lattice filters are commonly used in adaptive signal processing. One stage of a FIR lattice filter is shown in Fig. 9-3.

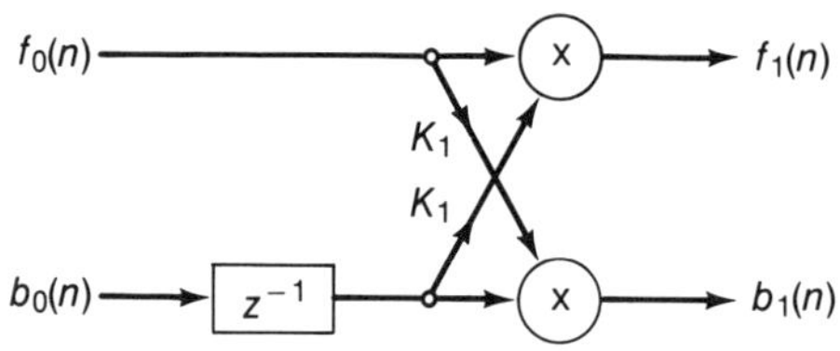

Figure 9-3. One stage of a lattice filter.

An Mth order lattice filter is obtained by connecting M stages in cascade (see Fig. 9-4).

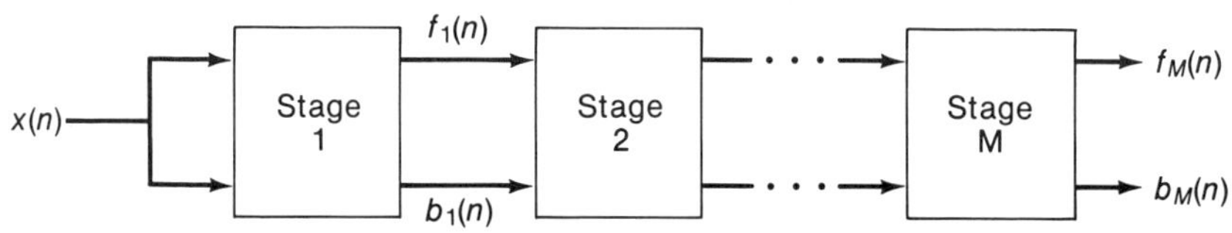

Figure 9-4 M-stage lattice filter.

Source listing of the macro file LATTICE.SRC is given in Fig. 9-5, and the child process LATTICE.FOR is listed in Fig. 9-6. Pay attention to the PARFILE statement in the macro file. This statement causes the macro registers of PC-DSP to be saved into a data file with a specified name (in this case, PARMS.DAT). Macro registers are saved in the following order:

String register 1
Floating-point register 1 (real part)
Floating-point register 1 (imaginary part)
Integer register 1
String register 2
.
.
.

Floating-point register 5 (imaginary part)

Integer register 5

Thus, a total of 20 items are written into the parameter file. It is the responsibility of the child process to position the file pointer at desired points in the file and read the registers of interest.

```
WINDOW  12  3  22  78
GETSTR  14  6  Output_:__   1
GETSTR  16  6  Input__:__   2
GETSTR  18  6  Coeffs_:__   3
GETINT  20  6  Number_of_stages_:__   1
ASAVE   @S2   TRANSFER.DAT
ASAVE   @S3   COEFFS.DAT
PARFILE  PARMS.DAT
OS-SHELL  LATTICE
```

Figure 9-5. Source code for LATTICE.SRC.

9-6 EXERCISES

1. Write a macro file to compute the log-magnitude of the DFT for a given sequence. The program should prompt the user for the names of input and output sequences.
2. Write a program to compute the Akaike information criterion (AIC) from the driving noise variance estimates of a sequence. Interface your program to PC-DSP through a macro file. The program should prompt the user for the name of the variable that holds the driving noise variance estimates. It should also ask for the name of the variable that will hold the AIC(M) sequence. It should then read the input from the file TRANSFER.DAT and write the output back into the same file.
3. Repeat Exercise 3 for the final prediction error (FPE) criterion defined by

$$\text{FPE}(M) = \sigma_M^2 \left(\frac{N+M+1}{N-M-1} \right)$$

4. Write a program to quantize a sequence with a specified number of quantization levels. Interface your program to PC-DSP with a macro file. Pass the number of quantization levels to your program through a parameter file.

```
C  LATTICE FOR
C
C  Program to implement a multi-stage lattice filter.
C  Written as an add-on to PC-DSP.
C
      DIMENSION X(1000), R(50)
      CHARACTER*30 DUMMY
C
C  Read the parameter file PARMS.DAT to get the number of
C  stages to be used.  The three READ statements immediately
C  following the OPEN statement are used for positioning the
C  file pointer to the integer register # 1.
C
      OPEN(UNIT=1, FILE='PARMS.DAT')
      READ(1, *) DUMMY
      READ(1, *) A
      READ(1, *) A
      READ(1, *) NSTG
      CLOSE(1)
C
C  Read the filter coefficients.
C
      CALL DATAIN2(R, M, 'COEFF.DAT')
C
C  Read the input data.
C
      CALL DATAIN2(X, NPTS, 'TRANSFER.DAT')
C
C  Call the lattice filter routine.
C
      CALL LATTFL(X, NPTS, NSTG, R)
C
C  Write the output sequence into TRANSFER.DAT.
C
      CALL DATAOU2(X, NPTS, 'TRANSFER.DAT')
      STOP
      END


      SUBROUTINE LATTFL(X, N, M, R)
C
C  Subroutine to implement a multi-stage lattice filter.
C
C  Input
C
C    X : Array that contains the input sequence.
C    N : Number of samples in X.
C    M : Number of lattice filter stages.
C    R : Lattice filter coefficients.
C
C  Output
C
C    X : Array that contains the output sequence.
C
      DIMENSION X(1), R(1), FW(1000), BW(1000)
C
C  Initialize.
C
```

Figure 9-6. Source code for LATTICE.FOR.

```
      DO 10 I=1,N
        FW(I) = X(I)
        BW(I) = X(I)
   10 CONTINUE
C
C   Start filtering.
C
      DO 20 I=1,M
        BPRV1 = 0
        DO 30 J=1,N
          BPRV2 = BW(J)
          BW(J) = R(I)*FW(J) + BPRV1
          FW(J) = FW(J) + R(I)*BPRV1
          BPRV1 = BPRV2
   30   CONTINUE
   20 CONTINUE
C
C   Put the result back into X.
C
      DO 40 I=1,N
        X(I) = FW(I)
   40 CONTINUE
      RETURN
      END


      SUBROUTINE DATAIN2(X, NPTS, FNAME)
C
C  This is a slightly modified version of the subroutine
C  DATAIN. It takes a third argument, FNAME, for the name
C  of the data file to be read.
C
      DIMENSION X(1)
      CHARACTER*30 FNAME
      OPEN(UNIT=1, FILE=FNAME, STATUS='OLD')
      I = 0
   10 I = I + 1
      READ(1, *, END=20) X(I)
      GOTO 10
   20 NPTS = I - 1
      CLOSE(1)
      RETURN
      END

      SUBROUTINE DATAOU2(X, NPTS, FNAME)
C
C  This is a slightly modified version of the subroutine
C  DATAOU. It takes a third argument, FNAME, for the name
C  of the data file to be read.
C
      DIMENSION X(1)
      CHARACTER*30, FNAME
      OPEN(UNIT=1, FILE=FNAME)
      DO 10 I=1, NPTS
      WRITE(1, *) X(I)
   10 CONTINUE
      RETURN
      END
```

Figure 9-6. (Cont'd.)

A

Function Reference List

This appendix lists PC-DSP functions in alphabetical order, along with a brief description of each. Mathematical algorithms used by the program, and the restrictions imposed on parameter values are detailed where applicable.

Add

Submenu containing two functions, *Seq+seq* and *Con+seq*, used for sample by sample addition of two sequences and addition of a constant to each sample of a sequence respectively. Located in the *Arithmetic* submenu of the *Data* component.

Analyze (FIR-filters)

Frequency-domain analysis of FIR filters. Located in the *FIR-filters* submenu of the *Filter* component. The variable containing the impulse response of the filter to be analyzed and the desired analysis type are specified by the user. The impulse response can be up to 512 samples long. Available analysis types are magnitude, log-magnitude, and phase. Two variables are generated, one holding the values for the desired analysis type, and one for the normalized frequency axis. PC-DSP uses FFT techniques for analysis of FIR filters.

Analyze (IIR-filters)

Frequency-domain analysis of IIR filters. Located in the *IIR-filters* submenu of the *Filter* component. The variable containing the description of the filter to be analyzed and the desired analysis type are specified by the user. Available analysis types are magnitude, log-magnitude, phase, group-delay, poles and zeros (tabular form or plot), second-order cascade section coefficients, and impulse response. For magnitude, log-magnitude, phase, and group-delay, two variables are generated, one holding the values for the desired analysis type and one for the normalized frequency axis. For impulse response, one variable is generated with the first 100 samples of the result.

Arithmetic

Submenu containing basic arithmetic functions on sequences. Located in the *Data* component. Contents are: *Add, Multiply, Conjg, Real, Imag, Even, Odd.*

aRma

Autoregressive moving average spectral analysis. Located in the *sPectrum* component. The input sequence, orders of numerator and denominator polynomials, and the desired analysis type are specified by the user. Available analysis types are PSD estimate (linear or logarithmic) and ARMA model coefficients. PC-DSP uses an ad hoc method for estimating the coefficients. First, the AR coefficients are estimated using the Yule-Walker method on the input data. Then the input data are filtered using the AR coef ficients, and a residual sequence is obtained. The MA coefficients are obtained from the residual sequence by the use of Durbin's method. (Large AR model order is chosen equal to twice the numerator order.) The maximum number of input samples is 1024, and the maximum ARMA model order is 50.

Auto-corr

Computes the autocorrelation estimate for a given sequence. Located in the *Process* submenu of the *Data* component. Uses the biased estimation formula:

$$\mathrm{Var}_1(m) = \frac{1}{N}\sum_{m} \mathrm{Var}_2^*(n)\mathrm{Var}_2(n+m) \qquad \text{for } m = 0, \ldots, N-1$$

The maximum length of the input sequence is 2048. PC-DSP uses FFT techniques for fast computation of the autocorrelation estimate.

Auto-Regr

Autoregressive (AR) spectral analysis. Located in the *sPectrum* component. The name of the input variable, AR model order, the desired method, and the desired analysis type are specified by the user. Available methods are the Yule-Walker method and the

Burg method. Biased autocorrelation lag estimates are used for both methods. Available analysis types are: AR coefficients, reflection coefficients up to a specified order, PSD estimate (linear or logarithmic), and driving noise variance estimates up to a specified order. The maximum number of input samples is 1024, and the maximum AR model order is 100.

Blackman

Generates a Blackman window sequence with a specified number of samples. The calling sequence is *Data / Generate / Window / Blackman.*

$$\text{Var}_1(n) = 0.42 - 0.5\cos(a) + 0.08\cos(2a) \text{ for } n = 0, \ldots, N-1$$

where

$$a = \frac{2\pi n}{N-1}$$

Maximum number of samples that can be generated is 4096.

Black-Tukey

Computes the Blackman-Tukey PSD estimate for a given sequence. Located in the *sPectrum* component. The name of the input variable, the largest autocorrelation lag to be used, and the desired window function are specified by the user. Autocorrelation lag estimates are computed using the biased estimation formula. The maximum number of input samples is 1024.

Clear

Clears all variables in memory and resets PC-DSP to its initial state. Located in the *System* component.

Conjg

Returns the complex conjugate of a sequence. Located in the *Arithmetic* submenu of the *Data* component.

$$\text{Var}_1(n) = \{\text{Var}_2(n)\}^* \qquad \text{for } n = 0, \ldots, N-1$$

Convolve

Computes the linear convolution of two sequences specified by the user. PC-DSP uses FFT techniques for fast convolution. If the two input sequences have N_2 and N_3 samples, respectively, the computed output sequence contains $N_1 = N_2 + N_3 - 1$ samples. This number should not exceed 4096.

Con+seq

Adds a user-specified real or complex constant to each sample of a sequence. Calling sequence is *Data / Arithmetic / Add / Con+seq.*

$$\text{Var}_1(n) = \text{Var}_2(n) + \text{Const} \qquad \text{for } n = 0, \ldots, N-1$$

If both $\text{Var}_2(n)$ and Const are real, then the resulting sequence $\text{Var}_1(n)$ is also real. Otherwise it is complex.

Con*seq

Multiplies each sample of a sequence with a user-specified real or complex constant. Calling sequence is *Data / Arithmetic / Add / Con*seq.*

$$\text{Var}_1(n) = \text{Var}_2(n) * \text{Const} \qquad \text{for } n = 0, \ldots, N-1$$

If both $\text{Var}_2(n)$ and Const are real, then the resulting sequence $\text{Var}_1(n)$ is also real. Otherwise it is complex.

Copy

Copies a specified range of a sequence into a new sequence. Located in the *Edit* submenu of the *Data* component. The desired range (N_1, N_2) is specified by the user. If $N_1 < N_2$, then the output is computed as

$$\text{Var}_1(n) = \text{Var}_2(n + N_1) \qquad \text{for } n = 0, \ldots, N_2 - N_1$$

If $N_1 > N_2$ then copying is done in reverse order, i.e.,

$$\text{Var}_1(n) = \text{Var}_2(N_1 - n) \qquad \text{for } n = 0, \ldots, N_1 - N_2$$

Cosine (Compute)

Computes the complex cosine of a sequence. Located in the *Nonlinear* submenu of the *Data* component.

$$\text{Var}_1(n) = \cos\{\text{Var}_2(n)\} \qquad \text{for } n = 0, \ldots, N-1$$

The cosine of a complex number $\alpha + j\beta$ is computed as

$$\cos(\alpha + j\beta) = \cos(\alpha) \cdot \cos h(\beta) - j \sin(\alpha) \cdot \sin h(\beta)$$

Cosine (Generate)

Generates a cosine waveform with specified parameters. Located in the *Generate* submenu of the *Data* component.

$$\text{Var}_1(n) = A \cos\{Bn + C\} \qquad \text{for } n = 0, \ldots, N-1$$

A, B, and C are constants that may be real or complex valued. N is limited to 4096 samples.

cRoss-corr

Computes the cross correlation estimate for two sequences. Located in the *Process* submenu of the *Data* component. Uses the biased estimation formula.

$$\mathrm{Var}_1(m) = \frac{1}{N}\sum_m \mathrm{Var}_2^*(n)\mathrm{Var}_3(n+m) \qquad \text{for } m = 0, \ldots, N-1$$

The maximum length of each input sequence is 2048. PC-DSP uses FFT techniques for fast computation of the cross correlation estimate.

Data

One of the six major components of PC-DSP. Contains basic waveform synthesis, processing, and editing functions.

Decimate

Generates a sequence by decimating the input sequence with a user-specified decimation factor. Located in the *Process* submenu of the *Data* component.

$$\mathrm{Var}_1(n) = \mathrm{Var}_2(Mn) \qquad \text{for } n = 0, \ldots, \mathrm{int}[(N-1)/M]$$

Delete

Deletes a variable from the internal storage area of PC-DSP. Located in the *System* component.

Design (FIR-Filters)

FIR filter design. Located in the *FIR-filters* submenu of the *Filter* component. The design method, the type of the filter to be designed, the number of taps, and the critical frequencies are specified by the user. Available design methods are the Fourier series design method, Parks-McClellan algorithm, and the least squares design method. Available filter types depend on the design method chosen. Low-pass, high-pass, band-pass, band-reject filters, differentiators, and Hilbert transform filters can be designed using the Fourier series method. Multiband filters, differentiators, and Hilbert transform filters are available with Parks-McClellan algorithm. The least-squares method designs multiband filters only. The maximum number of taps is 100. A printer report can be generated.

Design (IIR-Filters)

IIR filter design. Located in the *IIR-filters* submenu of the *Filter* component. The type of approximation and the type of filter to be designed are specified by the user. Available approximation types are Butterworth, Chebyshev type-I, and Chebyshev type-II (inverse Chebyshev). Available filter types are low-pass, high-pass, band-pass, and band-reject. Passband and stopband tolerances can be entered on either linear or logarithmic scale. Filter order can either be selected by the user or determined by the program as the smallest integer to satisfy the requirements. A printer report can be generated containing the details of the design procedure. PC-DSP uses the bilinear transformation method to design IIR filters from analog prototype filters. The maximum filter order is 50.

DFT

Computes the discrete Fourier transform of a given sequence by direct implementation of the DFT definition in Eq.(4-1). Located in the *Transform* submenu of the *Data* component.

$$\mathrm{Var}_1(k) = \mathrm{DFT}\{\mathrm{Var}_2(n)\}$$

Display

Displays the current graph on the screen. Located in the *Graphics* component.

Edit

Submenu containing waveform editing functions. Contents are *Copy, Flip, Join,* and *Makecomp*. Located in the Data component.

Even

Computes the conjugate symmetric component of a given sequence. Located in the *Arithmetic* menu of the *Data* component.

$$\mathrm{Var}_1(n) = 0.5\{\mathrm{Var}_2(n) + \mathrm{Var}_2^*((N-n))\}, \qquad \text{for } n = 0, \ldots, N-1$$

where double parentheses indicate modulo-N operation. If $\mathrm{Var}_2(n)$ is a real sequence, then $\mathrm{Var}_1(n)$ corresponds to the even part.

Exp (Compute)

Computes the complex exponential of a sequence. Located in the *Nonlinear* submenu of the *Data* component.

$$\mathrm{Var}_1(n) = \exp\{\mathrm{Var}_2(n)\} \qquad \text{for } n = 0, \ldots, N-1$$

The complex exponential of a complex number $\alpha + j\beta$ is computed as

$$\exp(\alpha + j\beta) = \exp(\alpha)\cdot\cos(\beta) + j\exp(\alpha)\cdot\sin(\beta)$$

Exp (Generate)

Generates an exponential waveform with specified parameters. Located in the *Generate* submenu of the *Data* component.

$$\text{Var}_1(n) = A\exp\{Bn + C\} \qquad \text{for } n = 0, \ldots, N-1$$

A, B, and C are constants that may be real- or complex-valued. N is limited to 4096 samples.

External

Generates an IIR filter description variable from the numerator and the denominator coefficients of an IIR filter transfer function. Located in the *IIR-filters* submenu of the *Filter* component. The IIR description variable obtained from the *External* function can be used for analysis of that filter.

Flip

Reverses the order of samples in a given sequence. Located in the *Edit* submenu of the *Data* component.

$$\text{Var}_1(n) = \text{Var}_2(N - 1 - n) \qquad \text{for } n = 0, \ldots, N-1$$

FFT

Computes the fast Fourier transform of a given sequence. Located in the *Transform* submenu of the *Data* component. Uses radix-2 decimation-in-time.

$$\text{Var}_1(k) = \text{FFT}\{\text{Var}_2(n)\}$$

If the length of $\text{Var}_2(n)$ is not an integer power of 2, it is internally zero-padded to the nearest integer power of 2.

Filter

One of the six major components of PC-DSP. Contains routines for design, analysis, and implementation of FIR and IIR filters.

FIR-filters

Submenu containing design and analysis routines for FIR filters. Located in the *Filter* component.

Generate

Submenu containing waveform generation functions. Located in the *Data* component. Contents are: *Ramp, Window, Impulse, Sine, Cosine, Exp, Normal, Uniform.*

Graphics

One of the six major components of PC-DSP. Used for generating customized graphs with user-defined mode, title, axis and data labels, and scaling options.

Hamming

Generates a Hamming window sequence with a specified number of samples. The calling sequence is *Data / Generate / Window / Hamming*.

$$\text{Var}_1(n) = 0.54 - 0.46 \cos\left(\frac{2\pi n}{N-1}\right) \qquad \text{for } n = 0, \ldots, N-1$$

Maximum number of samples that can be generated is 4096.

hAnning

Generates a Hanning window sequence with a user-specified number of samples. The calling sequence is *Data / Generate / Window / hAnning*.

$$\text{Var}_1(n) = \text{Imag}\{\text{Var}_2(n)\} \qquad n = 0, \ldots, N-1$$

Maximum number of samples that can be generated is 4096.

IIR-filters

Submenu containing design, analysis, and implementation routines for IIR filters. Located in the *Filter* component.

Imag

Returns the imaginary part of a complex sequence. Located in the *Arithmetic* submenu of the *Data* component.

$$\text{Var}_1(n) = \text{Imag}\{\text{Var}_2(n)\} \qquad n = 0, \ldots, N-1$$

Impulse

Generates a specified number of samples from a unit impulse sequence. Located in the *Generate* submenu of the *Data* component.

$$\text{Var}_1(n) = \delta(n) \qquad \text{for } n = 0, \ldots, N-1$$

N is limited to 4096 samples.

iNv-DFT

Computes the inverse discrete Fourier transform of a given sequence by direct implementation of the inverse DFT summation given by Eq.(4-2). Located in the *Transform* submenu of the *Data* component.

$$\text{Var}_1(n) = \text{DFT}^{-1}\{\text{Var}_2(k)\}$$

Inv-FFT

Computes the inverse fast Fourier transform of a given sequence. Located in the *Transform* submenu of the Data component. Uses radix-2 decimation-in-time.

$$\text{Var}_1(n) = \text{FFT}^{-1}\{\text{Var}_2(k)\}$$

If the length of $\text{Var}_2(k)$ is not an integer power of 2, it is internally zero-padded to the nearest integer power of 2.

inv-Z

Computes the inverse z-transform using long-division technique. Located in the *Transform* submenu of the *Data* component. Numerator and denominator coefficients and the desired number of inverse transform samples are specified by the user.

Join

Generates a sequence by joining two sequences, one appended to the end of the other. Located in the *Edit* submenu of the *Data* component.

$$\text{Var}_1(n) = \begin{cases} \text{Var}_2(n) & \text{for } n = 0, \ldots, N_2 - 1 \\ \text{Var}_3(n - N_2) & \text{for } n = N_2, \ldots, N_3 - 1 \end{cases}$$

Load (Graphics)

Loads a graph previously saved into a disk file. Located in the *Graphics* component. The loaded graph can be viewed by selecting *Display*.

Load (System)

Loads a sequence previously saved into a disk file. Located in the *System* component. Loaded files can be in either DSP or ASCII format. If an ASCII file is loaded, the name and the type of the loaded sequence is specified by the user.

Logarithm

Computes the natural logarithm of a complex sequence. Located in the *Nonlinear* submenu of the *Data* component.

$$\text{Var}_1(n) = \log\{\text{Var}_2(n)\} \qquad \text{for } n = 0, \ldots, N - 1$$

The logarithm of a complex number $re^{j\beta}$ is computed as

$$\log(re^{j\beta}) = \log(r) + j\beta$$

The imaginary part is the principal component of the phase. If this function is used for computing the complex cepstrum, phase unwrapping must be done by the user.

Macro

One of the six major components of PC-DSP. Used for executing macro files created using the macro compiler DSPMAC.

Magnitude

Computes magnitude of a complex sequence. Located in the *Nonlinear* submenu of the *Data* component.

$$\text{Var}_1(n) = |\text{Var}_2(n)| \qquad \text{for } n = 0, \ldots, N-1$$

Makecomp

Makes a complex sequence out of two real sequences. Located in the *Edit* submenu of the *Data* component.

$$\text{Var}_1(n) = \text{Var}_2(n) + j\text{Var}_3(n)$$

If $\text{Var}_2(n)$ and $\text{Var}_3(n)$ are of unequal lengths, the shorter sequence is automatically zero-padded to make the two sequences compatible.

Moving-Avg

Moving-average (MA) spectral analysis. Located in the *sPectrum* component. The name of the input sequence, MA model order, and the desired analysis type are specified by the user. Available analysis types are MA coefficients and the PSD estimate. Durbin's method is used for the estimation of MA model coefficients. The maximum number of input samples is 1024, and the maximum MA model order is 50.

Multiply

Submenu containing two functions, *Seq*seq* and *Con*seq*, used for sample-by-sample multiplication of two sequences and multiplication of each sample of a sequence by a specified constant, respectively. Located in the *Arithmetic* submenu of the *Data* component.

Nonlinear

Submenu containing nonlinear functions. Located in the *Data* component. Contents are: *Magnitude, Phase, Sine, Cosine, Tangent, Exp, Logarithm, sQ-root, Recipr.*

Normal

Normal random sequence generator. Located in the *Generate* submenu of the *Data* component. Generates a specified number of normally distributed random samples with specified mean and variance. Uses the central limit theorem. Number of samples is limited to 4096.

Odd

Computes the conjugate antisymmetric component of a given sequence. Located in the *Arithmetic* menu of the *Data* component.

$$\mathrm{Var}_1(n) = 0.5\{\mathrm{Var}_2(n) - \mathrm{Var}_2^*((N-n))\} \qquad \text{for } n = 0, \ldots, N-1$$

where double parentheses indicate modulo-N operation. If $\mathrm{Var}_2(n)$ is a real sequence, then $\mathrm{Var}_1(n)$ corresponds to the odd part.

Options

Lets the user modify the graph parameters. Located in the *Graphics* component. Graph mode, title, axis labels, axis ranges, data label increments and formats are specified by the user. A grid can also be specified.

OS-shell

Temporarily exits to the disk operating system. Located in the *System* component.

Periodogram

Computes the PSD estimate using Welch's method of averaging modified periodograms. Located in the *sPectrum* component. The name of the input variable, segment length, segment overlap, and the desired window function are specified by the user.

Phase

Computes phase of a complex sequence. Located in the *Nonlinear* submenu of the *Data* component.

$$\mathrm{Var}_1(n) = \mathrm{Phase}\{\mathrm{Var}_2(n)\} \qquad \text{for } n = 0, \ldots, N-1$$

Computed phase values are four-quadrant angles in radians, i.e.,

$$-\pi < \mathrm{Var}_1(n) \leq \pi$$

Print

Tabulates samples of a sequence either on the screen or on the printer. Located in the System component.

Process (Data)

Submenu containing functions to perform basic signal operations. Located in the *Data* component. Contents are: *Convolve, Auto-corr, cRoss-corr, Shift, Decimate, Zero-ins, zero-Pad.*

Process (IIR-filters)

Computes the output sequence of an IIR filter for a given input sequence. Located in the *IIR-filters* submenu of the *Filter* component. The name of the IIR description variable and the name of the input sequence are specified by the user. PC-DSP uses a cascade implementation of the filter consisting of second- and first-order blocks.

Quit

Exits PC-DSP and returns control to the disk operating system after the user confirms the decision to quit.

Ramp

Generates a linear ramp sequence with specified number of samples, initial value, and increment. Located in the *Generate* submenu of the *Data* component.

$$\mathrm{Var}_1(n) = \mathrm{Init} + n \cdot \mathrm{Incr} \qquad \text{for } n = 0, \ldots, N-1$$

N is limited to 4096 samples.

Read

Used for entering sample values of a sequence through the keyboard. Located in the *System* component. Up to 60 real or complex valued samples can be entered into a sequence. Samples of a sequence are separated by semicolons. Real and imaginary parts of a complex sample are separated by a comma.

Real

Returns the real part of a complex sequence. Located in the *Arithmetic* submenu of the *Data* component.

$$\mathrm{Var}_1(n) = \mathrm{Real}\{\mathrm{Var}_2(n)\} \qquad \text{for } n = 0, \ldots, N-1$$

Recipr

Computes the reciprocal of a sequence. Located in the *Nonlinear* submenu of the *Data* component.

$$\mathrm{Var}_1(n) = \frac{1}{\mathrm{Var}_2(n)} \qquad \text{for } n = 0, \ldots, N-1$$

Rectangular

Generates a rectangular window sequence with a specified number of samples. The calling sequence is *Data / Generate / Window / Rectangular*.

$$\mathrm{Var}_1(n) = 1 \qquad \text{for } n = 0, \ldots, N-1$$

Maximum number of samples that can be generated is 4096.

Reset

Resets graphics parameters to their default values. Located in the *Graphics* component.

Save (Graphics)

Saves a screen plot into a disk file. Located in the *Graphics* component. The plot must be displayed before it can be saved.

Save (System)

Saves a sequence into a disk file. Located in the *System* component. Can save sequences in either DSP or ASCII format.

Seq+seq

Sample by sample addition of two sequences. The calling sequence is *Data / Arithmetic / Add / Seq+seq*.

$$\mathrm{Var}_1(n) = \mathrm{Var}_2(n) + \mathrm{Var}_3(n)$$

If the sequences $\mathrm{Var}_2(n)$ and $\mathrm{Var}_3(n)$ are of unequal lengths, the shorter sequence is internally zero-padded to make the two sequences compatible.

Seq*seq

Sample-by-sample multiplication of two sequences. The calling sequence is *Data / Arithmetic / Add / Seq*seq*.

$$\mathrm{Var}_1(n) = \mathrm{Var}_2(n)\mathrm{Var}_3(n)$$

If the sequences $\text{Var}_2(n)$ and $\text{Var}_3(n)$ are of unequal lengths, the shorter sequence is internally zero-padded to make the two sequences compatible.

Shift

Shifts (delays) a sequence by a desired number of samples. Located in the *Process* submenu of the *Data* component.

$$\text{Var}_1(n) = \text{Var}_2(n - M)$$

Left shifts are also possible by using negative values of M. The permitted range of M is

$$-N_2 < M < 4096 - N_2$$

Sine (Compute)

Computes the complex sine of a sequence. Located in the *Nonlinear* submenu of the *Data* component.

$$\text{Var}_1(n) = \sin\{\text{Var}_2(n)\} \qquad \text{for } n = 0, \ldots, N-1$$

The sine of a complex number $\alpha + j\beta$ is computed as

$$\sin(\alpha + j\beta) = \sin(\alpha) \cdot \cos h(\beta) + j\cos(\alpha) \cdot \sin h(\beta)$$

Sine (Generate)

Generates a sine waveform with specified parameters. Located in the *Generate* submenu of the *Data* component.

$$\text{Var}_1(n) = A\sin\{Bn + C\} \qquad \text{for } n = 0, \ldots, N-1$$

A, B, and C are constants that may be real- or complex-valued. N is limited to 4096 samples.

sPectrum

One of the six major components of PC-DSP. Contains spectrum analysis functions.

sQ-root

Computes the square root of a complex sequence. Located in the *Nonlinear* submenu of the *Data* component.

$$\text{Var}_1(n) = \sqrt{\text{Var}_2(n)} \qquad \text{for } n = 0, \ldots, N-1$$

The square root of a complex number $re^{j\beta}$ is computed as

$$\sqrt{re^{j\beta}} = \sqrt{r} \cdot e^{j\beta/2}$$

System

One of the six major components of PC-DSP. Contains functions for loading and saving disk data files, keyboard entry of short sequences, listing, printing, and deleting variables in memory, and temporary exits to the disk operating system.

Tangent (Compute)

Computes the complex tangent of a sequence. Located in the *Nonlinear* submenu of the *Data* component.

$$\mathrm{Var}_1(n) = \tan\{\mathrm{Var}_2(n)\} \qquad \text{for } n = 0, \ldots, N-1$$

The tangent of a complex number $\alpha + j\beta$ is computed as

$$\tan(\alpha + j\beta) = \frac{\sin(\alpha + j\beta)}{\cos(\alpha + j\beta)}$$

Transform

Submenu containing transform functions. Located in the *Data* component. Contents are: *FFT, Inv-FFT, DFT, iNv-DFT, inv-Z.*

Triangular

Generates a triangular window sequence with a specified number of samples. Calling sequence is *Data / Generate / Window / Triangular.*

$$\mathrm{Var}_1(n) = \frac{|n-M|}{M} \qquad \text{for } n = 0, \ldots, N-1$$

where

$$M = \frac{N-1}{2}$$

N is limited to 4096 samples.

Uniform

Uniform random sequence generator. Located in the *Generate* submenu of the *Data* component. Generates a specified number of uniformly distributed random samples with specified lower and upper limits. Uses the multiplicative congruential method. The number of samples is limited to 4096.

Variables

Lets the user enter the names of the variables to be graphed. Located in the *Graphics* component. One x-axis variable and two y-axis variables can be entered. Only the first y- axis variable is required for a graph.

Var-list

Generates a screen listing of the variables in memory. Located in the *System* component.

Window

Submenu containing functions to generate window sequences. Located in the *Generate* submenu of the *Data* component. Contents are: *Rectangular, Triangular, Blackman, Hamming, hAnning.*

Zero-ins

Inserts zeros between samples of a sequence. Located in the *Process* submenu of the *Data* component. The zero insertion factor M is specified by the user. $M - 1$ zeros are inserted after each sample of the input sequence.

zero-Pad

Appends a specified number of zeros to the end of a sequence. Located in the *Process* submenu of the *Data* component.

B

Using PC-DSP in a Laboratory Environment

The purpose of this appendix is twofold: First, some general information will be provided on how PC-DSP was developed. This should be of particular interest to faculty members and students planning to develop educational software. Then, possible ways to utilize PC-DSP in a laboratory setting will be discussed, using the current laboratory implementation at Southern Illinois University—Edwardsville as an example. This part should be useful for instructors.

PC-DSP was developed using the compiled BASIC language. The program consists of over 8600 lines of code broken into 12 modules that are compiled separately and then linked together, resulting in the executable file PCDSP.EXE. The selection of compiled BASIC language for developing such a large program comes as a surprise to most people, but most objections to BASIC can be traced back to the early "interpreted" versions of the language that required line numbers and did not allow structure. Current implementations of the language do not require line numbers. Separately compiled subroutines with argument lists and local variables (very similar to FORTRAN subroutines) can be developed. Global variables can be declared through statements such as COMMON and SHARED. Rather sophisticated looping structures are available, similar to those found in the PASCAL language. Furthermore, very few other languages can match the ease and flexibility of screen control, graphics, and error recovery routines in compiled BASIC.

In the development PC-DSP, one of the guiding principles was to make the program as modular as possible, so that modifications and enhancements could be made easily. The program contains about 175 subroutines, each designed to perform one specific task. The main program does not do anything other than calling subroutines in the proper order. With this approach, most minor modifications to the program involve changes in just one or a few subroutines. Programming errors are also easier to find and eliminate.

PC-DSP is currently in use at Southern Illinois University—Edwardsville in conjunction with a senior-level digital signal processing course and its related laboratory sessions. It also finds occasional use in two junior-level linear systems courses. The senior-level digital signal processing course, EE 436, consists of 4 hours of lecture and one laboratory session per week for a 10-week academic quarter. There are a total of eight organized laboratory sessions conducted in weeks 2 through 9 of the quarter. An introductory session is given in the first week, and a make-up session is given in the tenth week for students who have missed one or more laboratory sessions. The student is expected to spend about $2\frac{1}{2}$ hours in each session.

In the first two sessions, basic waveform synthesis and manipulation techniques are emphasized. Students are asked to generate a number of sequences beginning with a simple sinusoid and moving up to more complicated waveforms. Then basic signal operations such as addition and multiplication of discrete sequences are studied using the sequences generated. A good number of convolution problems are given to reinforce the concept of convolution. Problems on decimation and interpolation are also given in the first two laboratory sessions.

Laboratory sessions 3 and 4 are built around the frequency domain transforms. Relationships between the discrete Fourier transform and the z-transform are explored. Students are asked to verify several DFT theorems numerically. The effects of zero-padding on the DFT are explored. The use of FFT for fast convolution is illustrated along with possible pitfalls. Session 4 also covers sampling and data acquisition using commercially available hardware.

Sessions 5 and 6 treat FIR and IIR filters respectively. Students use PC-DSP to design, analyze, and simulate a number of digital filters. In sessions 7 and 8, real-time signal processing is emphasized. Digital filters designed in previous sessions are implemented on Motorola DSP56001 processor. Code generators developed as PC-DSP macro files are used for generating the assembly language code for real-time implementation of designed filters.

References

BOSE, N. K., *Digital Filters, Theory and Applications,* Elsevier Science Publishing Co., New York, 1975.

DEFATTA, D. J., LUCAS, J. G., and HODGKISS, W. S., *Digital Signal Processing, A System Design Approach,* John Wiley & Sons, Inc., New York, 1988.

JACKSON, L. B., *Digital Filters and Signal Processing,* 2d ed., Kluwer Academic Publishers, Boston, 1989.

JOHNSON, J. R., *Introduction to Digital Signal Processing,* Prentice Hall, Englewood Cliffs, N.J., 1989.

HAMMING, R. W., *Digital Filters,* Prentice Hall, Englewood Cliffs, N.J., 1989

KAY, S. M., *Modern Spectral Estimation,* Prentice Hall, Englewood Cliffs, N.J., 1988.

KUC, R., *Introduction to Digital Signal Processing,* McGraw-Hill, New York, 1988.

LUDEMAN, L. C., *Fundamentals of Digital Signal Processing,* Harper & Row Publishers, New York, 1986.

MARPLE, S. L., *Digital Spectral Analysis with Applications,* Prentice Hall, Englewood Cliffs, N.J., 1987.

OPPENHEIM, A. V., and SCHAFER, R. W., *Discrete-Time Signal Processing,* Prentice Hall, Englewood Cliffs, N.J., 1989.

OPPENHEIM, A. V., and WILLSKY, A. S., *Signals and Systems,* Prentice Hall, Englewood Cliffs, N.J., 1983.

PARKS, T. W., and BURRUS, C. S., *Digital Filter Design,* John Wiley & Sons Inc., New York, 1987.

PROAKIS, J. G., and MANOLAKIS, D. G., *Introduction to Digital Signal Processing,* Macmillan Publishing Co., New York, 1988.

ROBERTS, R. A., and MULLIS, C. T., *Digital Signal Processing,* Addison-Wesley, Reading, Mass., 1987.

STRUM, R. D., and KIRK, D. E., *First Principles of Discrete Systems and Digital Signal Processing,* Addison-Wesley, Reading, Mass., 1988.

WELCH, P. D., The Use of Fast Fourier Transform for Estimation of Power Spectra, *IEEE Trans. Audio Electroucoust.* Vol. AU15, pp. 70–73, June 1967.

Index

PC-DSP 5 1/4" IBM® VERSION
By Oktay Alkin

Prentice-Hall

YOU SHOULD CAREFULLY READ THE FOLLOWING TERMS AND CONDITIONS BEFORE OPENING THIS DISKETTE PACKAGE. OPENING THIS DISKETTE PACKAGE INDICATES YOUR ACCEPTANCE OF THESE TERMS AND CONDITIONS. IF YOU DO NOT AGREE WITH THEM, YOU SHOULD PROMPTLY RETURN THE PACKAGE UNOPENED, AND YOUR MONEY WILL BE REFUNDED.

IT IS A VIOLATION OF COPYRIGHT LAW TO MAKE A COPY OF THE ACCOMPANYING SOFTWARE EXCEPT FOR BACKUP PURPOSES TO GUARD AGAINST ACCIDENTAL LOSS OR DAMAGE.

Prentice-Hall, Inc. provides this program and licenses its use. You assume responsibility for the selection of the program to achieve your intended results, and for the installation, use, and results obtained from the program. This license extends only to use of the program in the United States or countries in which the program is marketed by duly authorized distributors.

LICENSE

You may:

a. use the program;

b. modify the program and/or merge it into another program in support of your use of the program.

LIMITED WARRANTY

THE PROGRAM IS PROVIDED "AS IS" WITHOUT WARRANTY OF ANY KIND, EITHER EXPRESSED OR IMPLIED, INCLUDING, BUT NOT LIMITED TO, THE IMPLIED WARRANTIES OF MERCHANTABILITY AND FITNESS FOR A PARTICULAR PURPOSE. THE ENTIRE RISK TO THE QUALITY AND PERFORMANCE OF THE PROGRAM IS WITH YOU. SHOULD THE PROGRAM PROVE DEFECTIVE, YOU (AND NOT PRENTICE-HALL, INC. OR ANY AUTHORIZED DISTRIBUTOR) ASSUME THE ENTIRE COST OF ALL NECESSARY SERVICING, REPAIR, OR CORRECTION.

SOME STATES DO NOT ALLOW THE EXCLUSION OF IMPLIED WARRANTIES, SO THE ABOVE EXCLUSION MAY NOT APPLY TO YOU. THIS WARRANTY GIVES YOU SPECIFIC LEGAL RIGHTS AND YOU MAY ALSO HAVE OTHER RIGHTS THAT VARY FROM STATE TO STATE.

Prentice-Hall, Inc. does not warrant that the functions contained in the program will meet your requirements or that the operation of the program will be uninterrupted or error free.

However, Prentice-Hall, Inc., warrants the diskette(s) on which the program is furnished to be free from defects in materials and workmanship under normal use for a period of ninety (90) days from the date of delivery to you as evidenced by a copy of your receipt.

LIMITATIONS OF REMEDIES

Prentice-Hall's entire liability and your exclusive remedy shall be:

1. the replacement of any diskette not meeting Prentice-Hall's "Limited Warranty" and that is returned to Prentice-Hall with a copy of your purchase order, or
2. if Prentice-Hall is unable to deliver a replacement diskette or cassette that is free of defects in materials or workmanship, you may terminate this Agreement by returning the program, and your money will be refunded.

IN NO EVENT WILL PRENTICE-HALL BE LIABLE TO YOU FOR ANY DAMAGES, INCLUDING ANY LOST PROFITS, LOST SAVINGS OR OTHER INCIDENTAL OR CONSEQUENTIAL DAMAGES ARISING OUT OF THE USE OR INABILITY TO USE SUCH PROGRAM EVEN IF PRENTICE-HALL OR AN AUTHORIZED DISTRIBUTOR HAS BEEN ADVISED OF THE POSSIBILITY OF SUCH DAMAGES, OR FOR ANY CLAIM BY AN OTHER PARTY.

SOME STATES DO NOT ALLOW THE LIMITATION OR EXCLUSION OF LIABILITY FOR INCIDENTAL OR CONSEQUENTIAL DAMAGES, SO THE ABOVE LIMITATION OR EXCLUSION MAY NOT APPLY TO YOU.

GENERAL

You may not sublicense, assign, or transfer the license or the program except as expressly provided in this Agreement. Any attempt otherwise to sublicense, assign, or transfer any of the rights, duties, or obligations hereunder is void.

This Agreement will be governed by the laws of the State of New York.

Should you have any question concerning this Agreement, you may contact Prentice-Hall, Inc., by writing to:

Prentice Hall
College Division
Englewood Cliffs, N.J. 07632

Should you have any questions concerning technical support you may write to:

Professor Oktay Alkin
Department of Electrical Engineering
Southern Illinois University @ Edwardsville
Edwardsville, IL 62026-1801

YOU ACKNOWLEDGE THAT YOU HAVE READ THIS AGREEMENT, UNDERSTAND IT, AND AGREE TO BE BOUND BY ITS TERMS AND CONDITIONS. YOU FURTHER AGREE THAT IT IS THE COMPLETE AND EXCLUSIVE STATEMENT OF THE AGREEMENT BETWEEN US THAT SUPERCEDES ANY PROPOSAL OR PRIOR AGREEMENT, ORAL OR WRITTEN, AND ANY OTHER COMMUNICATIONS BETWEEN US RELATING TO THE SUBJECT MATTER OF THIS AGREEMENT.

ISBN 0-13-655200-5